T0112769

Understanding Cancer

One in two of us will develop cancer at some point in our lives and yet many of us don't understand how cancers arise. How many different kinds of cancer are there? What treatments are available? What does the future hold in terms of developing new therapies?

This book demystifies cancer by explaining the underlying cell and molecular biology in a clear and accessible style. It answers the questions commonly asked about cancer, such as what causes cancer and how cancer develops. It explains how DNA makes proteins and how mutations can corrupt those proteins. It also gives an overview of current therapies and how treatments may advance over the next decades, as well as explaining what actions we can take to help prevent cancer developing.

Understanding Cancer is an accessible and engaging introduction to cancer biology for any interested reader.

Robin Hesketh has been a member of the Department of Biochemistry at the University of Cambridge and a fellow of Selwyn College for over 40 years, working on cancer biology. He has published over 100 research papers, a textbook on cancer (*Introduction to Cancer Biology*, Cambridge University Press, 2013) and popular science books (*Betrayed by Nature*, Palgrave, 2012). He has spoken and written widely in the media on cancer and has run a blog on the topic of cancer for the general public since 2011.

The **Understanding Life** series is for anyone wanting an engaging and concise way into a key biological topic. Offering a multidisciplinary perspective, these accessible guides address common misconceptions and misunderstandings in a thoughtful way to help stimulate debate and encourage a more in-depth understanding. Written by leading thinkers in each field, these books are for anyone wanting an expert overview that will enable clearer thinking on each topic.

Series Editor: Kostas Kampourakis http://kampourakis.com

Published titles:

Forthcoming:

Understanding Cancer

ROBIN HESKETH
University of Cambridge

CAMBRIDGE
UNIVERSITY PRESS

CAMBRIDGE
UNIVERSITY PRESS

University Printing House, Cambridge CB2 8BS, United Kingdom

One Liberty Plaza, 20th Floor, New York, NY 10006, USA

477 Williamstown Road, Port Melbourne, VIC 3207, Australia

314–321, 3rd Floor, Plot 3, Splendor Forum, Jasola District Centre,
New Delhi – 110025, India

103 Penang Road, #05–06/07, Visioncrest Commercial, Singapore 238467

Cambridge University Press is part of the University of Cambridge.

It furthers the University''s mission by disseminating knowledge in the pursuit of
education, learning, and research at the highest international levels of excellence.

www.cambridge.org
Information on this title: www.cambridge.org/9781316517178
DOI: 10.1017/9781009043243

First published 2022

Printed in the United Kingdom by TJ Books Limited, Padstow Cornwall

A catalogue record for this publication is available from the British Library.

Library of Congress Cataloging-in-Publication Data
Names: Hesketh, Robin, author.
Title: Understanding cancer / Robin Hesketh, University of Cambridge.
Description: Cambridge, United Kingdom ; New York, NY : Cambridge University
Press, [2022] | Series: Understanding life | Includes index.
Identifiers: LCCN 2021050288 (print) | LCCN 2021050289 (ebook) | ISBN
9781316517178 (hardback) | ISBN 9781009043243 (ebook)
Subjects: LCSH: Cancer – Popular works. | BISAC: SCIENCE / Life Sciences /
Developmental Biology
Classification: LCC RC263 .H475 2022 (print) | LCC RC263 (ebook) | DDC 616.99/4–
dc23/eng/20211028
LC record available at https://lccn.loc.gov/2021050288
LC ebook record available at https://lccn.loc.gov/2021050289

ISBN 978-1-316-51717-8 Hardback
ISBN 978-1-009-00599-9 Paperback

'How often have we attended a lecture or opened a book to find that within minutes we are smothered by complicated facts that are way beyond our understanding? There has been no simple introduction. The speaker/author is so involved in the topic that they could no longer see out of the intellectual hole that they had dug for themselves. If ever a book was written to dispel this fault, then this is the one, as Robin Hesketh has managed to provide a remarkably clear and readable account of the science behind cellular behaviour and faults that lead to the development of cancer. We become convinced that the key to cancer is DNA mutation with chopping and changing of DNA strands, and the older you are the more likely this is to occur. Read why the tumour suppression protein (p53) can make cells commit suicide but does not always work.

The book reads like a novel, and I found that I could hardly put it down. The literary style is at times light-hearted with humorous analogies.'

Robert Whitaker, Anatomist, University of Cambridge

'*Understanding Cancer* presents a carefully crafted, clear and concise book on aspects of cancer; a disease of importance to us all. Most readers will come to Robin Hesketh's book with questions about cancer. *Understanding Cancer* will not disappoint. The most usual questions and answers are presented in the first chapter and ways of reducing the risk of some cancers are suggested later.

This book puts cancer into a historical and very interesting context; it then explores cancer, its biochemistry and functioning in an approachable way. Information is given about the latest treatments and the science behind them. This very readable book contains something for everyone. It is positioned in, and very adequately fills, the gap between personal accounts by patients of their experiences, and more advanced medical and cell biology texts. *Understanding Cancer* is well researched and greatly recommended.'

David Archer, Schools Liaison Officer, British Society for Cell Biology

'*Understanding Cancer* is a fascinating and engaging perspective on the evolution of cancer research and treatment. Dr Hesketh provides insight into the key clinicians and scientists, following their discoveries in clinical care and research. It is clear that achievements in cancer treatments are rooted in basic research, and the book highlights the collaborations required between scientists and oncologists in order to make the next leap of advances in treating cancer. Dr Hesketh overviews the likely mutagenic causes of cancer spurring on the oncogenic transitions leading to a cancer cell that can replicate uncontrollably. He also highlights new avenues in cancer research, such as studies on components of the tumor microenvironment (i.e. blood vessel cells, immune cells), which can then lead to the development of additional 'ammunition' to battle cancer. Dr Hesketh conveys that preventive measures and advances in early cancer detection could make an impact on cancer incidence and patient outcomes/survival. This book is certainly a triumph and a must-read for all current and future scientists, physicians at any stage of their professional careers and anyone interested in cancer research and the quest for effective anti-cancer treatments.'

David Lyden, Cancer Researcher and Paediatric Oncologist,
Weill Cornell Medicine, Cornell University

Without the limitless love and support of my wife, Jane, and my two sons, Robert and Richard, I could not have written this book nor indeed done much else in my life.

Contents

Colour plates can be found between pages 200 and 201.

Foreword

What causes cancer? This is a question that many people ask. The answer? There is no simple one. The term "cancer" is commonly used to describe a variety of diseases that share certain features such as uncontrolled cell proliferation. During the last several decades numerous researchers have tried to reveal the causes of cancer through the study of mutations and their impact on phenotypes. The main underlying idea has been that mutations, either caused by accident or by environmental factors, are responsible for the uncontrolled proliferation of cells in which they occur. In the present book, Robin Hesketh provides a detailed and informative account of what we know, as well as what we do not know, about the impact of mutations with respect to cancer. Readers will find a wealth of information, explained in a concise and clear manner.

Kostas Kampourakis, Series Editor

Preface

Cancer is unique. No other human condition is quite so two-faced in being, on the one hand, easy to grasp in terms of its basic cause, while, on the other hand, confronting us with overwhelming complexity when we get down to details. Highly appropriate, then, that if you delve into the huge toolbox of cancer drivers you'll find a gene called *Janus* – after the mythological god of beginnings and transitions, usually depicted as having two faces, one looking to the past and the other towards the future.

Cancer's past is immense: we know it has afflicted animals for millions of years. However, a cancer biologist might argue that some 60-odd years ago it turned its face to the future. The critical event was, of course, the revelation of the structure of DNA and all that followed. Henceforth cancer biology became the science of molecules – genes and proteins and how they cause cancers – that continues to build the foundations of an understanding of cancer and rational approaches to therapy.

If this sounds as though I am casting aside the heroic efforts of numerous great scientists and physicians who, starting with the ancient Egyptians and Chinese, attempted to grapple with the cancer challenge, rest assured that this is not the case, and Chapter 2 surveys the major events that preceded the age of molecular biology.

After that we look at cancer numbers worldwide and how their sheer scale begins to tell us something about underlying causes. Then, on to DNA and how the code it carries was worked out and the picture we now have of cells turning its message into a limitless number of proteins that define each species and enable all living things to function. Chapter 5 reveals how cells work – in particular how one cell becomes two – and that leads to how disruptions in DNA compromise the delicate machinery of replication to give rise to cancers.

We then turn to the causes of cancer, familiar – tobacco, alcohol, etc. – and less well-known – bacteria, fungi, etc. – and, most importantly, what, if anything, can be done about them.

Cancers may be treated by surgery, radiotherapy or drugs (chemotherapy), often in combinations. Chapter 9 considers the current state of play with the emphasis on chemotherapy and molecular approaches to treatment. This leads to the final chapter which looks to the future by reviewing the astonishing range of innovative strategies that are under development. Some of these are relatively advanced (e.g., immunotherapy), others are embryonic. Some will fall by the wayside but, collectively, they represent extraordinary science and offer great promise that, after so many millennia, mankind may at last be able to control these dreaded yet fascinating diseases.

We begin, however, by answering a dozen or so of the most likely questions a newcomer to cancer might ask or indeed that children often do ask. This should clear up misunderstandings that are common in the general perception of cancer and set the stage for the exciting voyage to come.

Acknowledgements

In writing a story of cancer for non-specialists I owe a massive amount to countless people I've been fortunate to meet in the course of my career. Scientific colleagues, clinicians, patients, students, and members of the public who've been kind enough to come to my talks or read my blogs and books – so many that I've been privileged to encounter and to learn from. Enormous thanks to Katrina Halliday of Cambridge University Press, without whom this book would not have happened, and also to Jessica Papworth and Kostas Kampourakis for brilliant editing. Many thanks also to my colleague Thomas Shafee who drew the originals for several figures. I am also very appreciative of the work of Olivia Boult, Sam Fearnley, Gary Smith, Judith Reading, Gayathri Tamilselvan and Vigneswaran Viswanathan in the production stages.

Gene Names

The HUGO Gene Nomenclature Committee (HGNC: www.genenames.org /index.html) assigns unique symbols to human genes. Gene names are written in italicized capitals: the protein that they encode is non-italicized: *EGFR* (gene)/EGFR (protein). They are pronounced phonetically when possible (SRC is *sarc*, MYC is *mick*, ABL is *able*). Viral forms are prefixed by v- (e.g., v-*src*). For some genes that have commonly used informative names both are shown (e.g., *SLC2A1*/GLUT1 and *SLC2A5*/ GLUT3).

Chemical and Trade Names of Drugs

Throughout the text chemical names of drugs are used. This list provides corresponding trade names.

> 5-fluorouracil (Adrucil and others)
> Azacitidine (Vidaza)
> Cetuximab (Erbitux)
> Docetaxel (Taxotere and others)
> Enasidenib (Idhifa)
> Erlotinib (Tarceva)
> Fulvestrant (Faslodex and others)
> Gefitinib (Iressa)
> Gemcitabine (Gemzar)
> Imatinib (Gleevec)
> Larotrectinib (Vitrakvi)

Methotrexate, formerly amethopterin (Trexall, Rheumatrex, Otrexup and others)

Olaparib (Lynparza)

Paclitaxel (Taxol and others)

Palbociclib (Ibrance and others)

Pembrolizumab, formerly lambrolizumab (Keytruda)

Raloxifene (Evista and others)

Rituximab (Rituxan and others)

Tamoxifen (Nolvadex and others)

Toremifene (Fareston)

Trastuzumab (Herceptin)

Vemurafenib (Zelboraf)

1   Painting a Clear Picture

Once upon a time it was fair to say that most people knew little of science. After all, scientists spent years learning their job so it's clearly tough-going and, by and large, the rest of the world could get by knowing nothing of superconductivity or the origins of the universe. But increasingly our daily lives have come to be dominated by science, and part of that revolution has been the ever-expanding reach of television and the Internet as sources of information. It's as though, unwittingly, we've all signed up to the Open University. And, it should be said, when it comes to science this has all been helped by a growing awareness among those in the trade that they have an obligation to let the world know how they while away their days.

High time too, I say – but all might not agree. Friends of mine who are general practitioners tell me that increasingly folk turn up to their surgeries fully armed with a diagnosis of their perceived condition, courtesy of the Internet. I can see that being forced to wonder why you spent years slogging through medical school could be rather dispiriting but, as I point out, setting the personal aside, you have to concede that people being willing and able to teach themselves has to be a step forwards for civilization.

That's as may be, but in science, and especially in biology, there is almost always another side to any argument. Alexander Pope, in his 1709 essay, noted that 'A little learning is a dangerous thing', and that 'shallow draughts intoxicate the brain'. Wise words, and with them in mind let's make a start on understanding cancer by looking at some critical questions that often cause confusion – and not just for non-scientists!

As we launch ourselves into this story, there's one thing we all might agree on: cancer is complicated. However, and you might find this surprising, what emerges from short answers to a handful of rather obvious questions is that cancer is a great paradox. On the one hand, it is indeed mind-boggling – which is why clinicians will almost never say 'This will work' with regard to treatments. But, when you cut to the heart of the matter, cancer is very simple – that is, it's easy to grasp the key points and thus to see how, in principle, to go about dealing with it. It's only then that the going gets tough as our ingenuity is tested by the immense weight of evolution that underlies the disease.

What is Cancer?

A three-word answer to this most basic of questions is 'Cells behaving badly'. More scientifically, it's a group of cells somewhere within an animal that are reproducing (i.e., making more of themselves) either faster than they should or in a place where they should not be. Put another way, these cells have lost control of their capacity to divide. The result is that the cells grow and divide to make more copies of themselves, paying no heed to normal controls. The resulting unruly mass of cells constitutes a tumour. This word comes from the Latin for 'swelling' – that is, an abnormal growth. It's used interchangeably with 'cancer' and 'neoplasm' (new growth), and all three words mean much the same. Tumours may grow relatively quickly or very slowly, but to expand significantly food and oxygen are required – just like for any other cell in the animal body. To achieve growth, tumours can release chemical signals that switch on growth in nearby blood vessels. New 'sprouts' penetrate into the tumour cell mass. The whole ensemble is now primed to take off: to expand regardless of the best interests of its host. Most ominously of all, by acquiring its own blood supply the tumour now has a conduit and cancer cells can be carried the circulatory system to any part of the body. Cells may also be carried by the lymphatic system but, whatever the means of transport. The process of tumour cells spreading to other places is called metastasis and it's critical because it results in over 90 per cent of deaths due to cancer.

What Causes Cancer?

Perhaps the most frequently asked question about cancer. In today's world most of us can come up with a quick answer: mutations – that is, damage to our genetic material, otherwise known as DNA. The term DNA has, of course, passed into common speech as we have all learned more and more about the science of molecular biology. That's good, because it means we don't need to use its full name (deoxyribonucleic acid) when, in Chapter 4, we come back to this wondrous molecule that is the eternal language of the living cell and look at how it works and the many ways in which it can be 'damaged'. The key point for now is that what we inherit comes in the form of DNA – a gigantic chemical molecule made of four types of unit joined together. The units are bases (abbreviated as A, C, G and T), and it is their sequence that encodes genetic information.

The term 'genome' was invented in 1920 to describe the entire genetic material of an organism – thus, in humans that includes the DNA in the nucleus of our cells (almost all our DNA) and a tiny amount in mitochondria, membrane-bound units in cells often called the powerhouse of the cell. It's mitochondria, and hence mitochondrial DNA, that are passed almost exclusively from mother to offspring in the egg.

Within the length of DNA there are blocks called genes, which carry specific, functional units of heredity. The study of genes and genetic variation is called genetics. The study of the genome is called genomics. Before you ask, the Danish botanist Wilhelm Johannsen is credited with coining the word 'gene' ('gen' in Danish and German) in 1909. Because cancers arise from changes in genetic material, they are 'genetic diseases'. That doesn't make them unique: about 6,000 other genetic disorders are known. What is unique about cancers is that almost all need mutations in several genes to get them started and keep them going. Thus, the vast majority of cancers arise from the combined effects of mutated genes. We should note that about 20 per cent of cancers are initiated by infection but, ultimately, these too acquire mutations that drive tumour development.

Given that they are genetic diseases, you might suppose that cancers could be passed from one generation to another through defective genes – and indeed they can. As long ago as 1820, a Stourbridge doctor, William Norris, described a family in which individuals from several generations had developed the same form of cancer. This inference that some families might be predisposed to cancer was extended by the extraordinary French physician Paul Broca who, in 1866, suggested it might be possible to inherit breast cancer. He'd looked at his wife's family tree and noted that 10 out of 24 women, spread over four generations, had died from that disease and that there had been cases of other types of cancer in the family as well. We know now, of course, that a changed (mutated) form of a gene passed from generation to generation was almost certainly responsible for the suffering of this family.

One consequence of the rise of the 'media' is that breast cancer genetics has in recent times come into the spotlight, with the much-publicized saga of Angelina Jolie, the American actress. Jolie's mother and maternal grandmother had both died of ovarian cancer, and her maternal aunt from breast cancer – a family history that persuaded Jolie to opt for genetic testing that indeed revealed she was carrying a mutation in one of two genes named *BRCA1* and *BRCA2* (the acronyms come from BReast CAncer type 1 and type 2, so named because they were the first major genes to be identified, in 1990 and 1994, as associated with hereditary breast cancer). *BRCA* genes are mutated in about 10 per cent of breast cancers and 15 per cent of ovarian cancers. The National Cancer Institute estimates that 'about 12% of women in the general population will develop breast cancer sometime during their lives'. By contrast, a recent large study estimated that about 72 per cent of females who inherit a harmful *BRCA1* mutation and about 69 per cent with a harmful *BRCA2* mutation will develop breast cancer by the age of 80.

These estimates prompted Jolie to have a preventive double mastectomy, thereby reducing her risk to less than 5 per cent. The 'Angelina effect' saw a doubling in the number of women being referred for genetic testing for breast cancer mutations in the months after she revealed her story. A study in 2020 concluded that screening entire populations for *BRCA* mutations, rather than only those with a strong family history of breast or ovarian cancer, could prevent millions of breast and ovarian cancer cases worldwide. For the UK the estimate was that about 10,000 deaths from these cancers would be prevented.

Breast cancers are an enormously varied set of diseases, and as such they're a challenge even to classify, let alone to treat. The recent rapid progress in DNA sequencing has led to a new genome-based classification system but there is still strong reliance on the traditional prognostic and predictive factors, notably what's called hormonal status – meaning the presence on the surface of the tumour cells of protein receptors to which the hormones oestrogen and progesterone attach, together with the presence or otherwise of the human epidermal growth factor receptor 2 (HER2). One significant sub-group has no detectable levels of these proteins. These are called triple-negative breast cancers (TNBCs), and they make up 10–15 per cent of breast cancers. They are very aggressive cancers (i.e., have a poor prognosis), known for some years to disproportionally affect young women of African origin – they are about twice as common in African Americans as in European Americans. Sequencing has revealed that mutations in *BRCA1* are present in most (69 per cent) TNBCs in females of European origin. But here's a very odd thing: African American women have a low *incidence* of *BRCA1* mutations (less than 20 per cent – incidence being the number of new cancers occurring in a population per year), despite the fact that they are relatively prone to TNBC. This implies, of course, that if *BRCA1* isn't doing the driving there must be other potent drivers for TNBC in this group.

These examples clearly show that cancer can 'run in families' and the estimate is that 10–30 per cent of cancers arise from inherited genetic damage. However, the majority occur as the result of accumulated DNA damage as we pass through life. In other words, cancers are, by and large, diseases of old age. In the UK and the USA about 70 per cent of all newly diagnosed cancers occur in people aged 60 or over. Knowing the rate at which we collect mutations, it's easy to work out that if we lived to be 140 years old we'd all have a cancer of some sort. 'Thank heavens we don't have to worry about that yet' is a perfectly reasonable reaction, but there's an important point here, namely that the fact of the inevitability of cancer (if we live long enough) tells us that it's an in-built feature of life. It may be difficult to deal with, but it's not something freaky and weird. It arises because our

DNA is not made of stone: it's mutable and hence vulnerable, as indeed it must be, for without its plasticity there would be no evolution.

Are All Cancers Equally Bad?

We've just noted that the critical event in terms of potential lethality is the acquisition of metastatic capacity: the tumour is no longer self-limiting in terms of growth, it can invade adjacent tissues and spread to distant sites. In short, it's become malignant. However, it may have occurred to you that if cancer cells have to do 'something' to become malignant, it is quite likely that many of them won't bother. Indeed, a lot of them do just that (nothing, that is) and we've known since early in the twentieth century that mini-tumours can form and then stop growing, remaining static as 'dormant tumours'. The most likely reason is that they are not able to flip the switch that turns on the growth of new blood vessels.

It has transpired from autopsies of road traffic accident victims that many, perhaps all, adults are wandering around carrying dormant tumours – clumps of about 100,000 cells – in a variety of organs and tissues. Sometimes called *in-situ* tumours, these microscopic growths would normally never be detected – it just happened that accidental deaths provided tissues for pathological analysis.

The key point here is that these micro-tumours were clearly dormant: their carriers died in accidents and had shown no signs of cancer. Knowing what we do about the time course of cancer development, we can be sure that most of them would not have gone on to produce cancer for many more years, or even decades.

Malignant versus Benign

We've now met the two ends of the cancer spectrum: dormant tumours that we can ignore and malignant tumours that we ignore at our peril. We should note in passing that malignancy is preceded by a pre-malignant phase, namely groups of cells (lesions) that are not yet cancerous but have the potential to develop into malignant cancer (i.e., become metastatic). One example would be colon polyps, growths on the lining of the colon or rectum that can progress to bowel cancer.

There's one further group of cancers that we need to meet – not least because almost all of us have got some of these too – benign tumours. They are indeed extremely common. For example, in 9 out of 10 women it's possible to detect changes in breast tissue that are benign and not dangerous. Fibroids are another type of abnormal growth: they occur in the uterus and are also typically benign. And that's the most important thing about benign tumours: they're not malignant – that is, they can't invade surrounding tissues and therefore do not spread. Benign tumours can arise in any tissue, the most common being lumps of fat called lipomas and, in general, they are fairly harmless. They're usually surrounded by a membrane, a sort of sac that helps to prevent them from spreading. They tend to grow very slowly, but they can reach the size of a grapefruit. The only real problem comes if they press on other tissues (e.g., blood vessels or the brain). That may require surgical treatment, but the good news is that once removed they usually don't return.

One other way in which they can have harmful, indirect effects is by growing in tissues that make hormones, such as the adrenal glands or the thyroid. When this happens, the tumours are derived from cells of the tissue and you might guess that the extra growth would give rise to abnormal levels of the hormones normally made by those glands. They're often symptom-free and only detected by chance (say, from a blood test).

An obvious thought is that, if the evolution of malignant cancers is driven by picking up changes in DNA, perhaps benign growths don't arise from mutations but are just caused by, say, a local imbalance in growth factors – chemical signals that turn on cell proliferation. As ever in cancer, it's not that simple. Mutations that in some tissues are associated with malignancy also pop up in benign tumours and in normal tissue, which tells us that knowing the mutational state of genes doesn't enable you to say for sure whether a growth will become malignant. We're stuck with what, as this story unfolds, you will come to recognize as a typical cancer problem. The difference between benign and malignant tumours is critical: one of them can kill you. But even with the all-conquering power of modern molecular biology that we will come to shortly, we are yet to define precisely what it is that converts a relatively harmless abnormal growth into the fatal variety.

Warts and All

I suspect everyone will have noticed that human beings tend to come adorned with a variety of moles, birthmarks and warts. Try as you might, it's hard not to ask yourself sometimes whether these things, that are undoubtedly unusual growths, are some form of cancer – and if they are, what should be done about them. Relax. The answers are almost always 'no' and 'nothing'. If you want to be pedantic, as abnormal growths of skin they are indeed 'neoplasms', but the best thing is to forget about them or, if they're Angelina Jolie's mole or Mikhail Gorbachev's port-wine stain, turn them into an adornment. Sometimes these oddities will disappear of their own accord, as often happens with 'strawberry marks' usually found on the face. Otherwise, if they are a cosmetic concern, it's often possible to reduce their prominence by laser treatment.

One other abnormality you may acquire is a cyst. These are not benign tumours but are closed sacs of cells containing liquid or semi-solid material that can form almost anywhere and can be removed by surgery.

You probably spotted another reference to the enigma of cancer a few lines ago in 'almost always', and, although we'll return to this point later, we need a word about moles before we get to warts. These are birthmarks, called nevi, the most common form being a growth of melanocytes in the outer layer of the skin. These cells make melanin – a skin pigment that gives a dark colour to hair, skin and eyes. Moles are therefore benign clusters of pigmented skin cells. Normally no more than decorative, just once in a while a mole can kick off into something nasty and turn into a fully malignant tumour. No need for panic, however, for that almost always requires us to give it a helping hand – usually by lying in the sun without any protection, that is, by exposing it to ultraviolet radiation. Hence the widely publicized advice to use sun cream. Regardless of sun or creams, the essential thing is to consult a doctor if one of your moles changes appearance – gets blacker, starts growing, itching or bleeding. At that point the problem can be resolved by surgical removal of the offending spot. We'll return to the other scenario later, when we look at drug treatments for malignant melanoma.

The key thing that distinguishes birthmarks and moles from warts is that warts are caused by viral infection. That means we aren't born with them, but most of us get them at some point, often before we are 20 years old. More often than not they disappear of their own accord, although they may take years to do so. Usually they form on the hands or feet or in the anogenital area. Palmar warts occur on the palm of the hand; plantar warts, otherwise known as verrucas (*verruca plantaris*), on the soles of the feet. It's worth noting that warts are not the same as the irritating condition known as athlete's foot (*tinea pedis*), which is a fungal infection of the skin between the toes.

Warts are caused by infection with human papillomavirus (HPV), which means they are contagious. There are over 100 different types of HPV, giving rise to variant forms of warts in the outer layer of skin (epidermis) where the virus causes excessive amounts of a protein called keratin to be made. Once infected, you can't get rid of HPV. Nevertheless, most warts can be treated either chemically or by freezing (cryosurgery), burning (cauterization) or laser treatment.

Why Do Some Children Get Cancer?

They're very unlucky. Either they've inherited a powerful cancer-driving mutation in the DNA they received from a parent, or they acquired such a mutation in the womb. When our very young are stricken it is, of course, especially shattering, so it's worth pointing out that childhood cancers are very rare – less than 1 per cent of all cancers. In the UK the yearly incidence is about 1 child in 500 under the age of 14 – around 1,900 in total with just over 200 deaths. The corresponding US figures are just over 11,000 new cases with 1,200 deaths. The most common childhood cancer affects the blood cells (acute lymphoblastic leukaemia). For this disease the wonderful advances of the last 50 years have seen the cure rate soar to 90 per cent from about 50 per cent in the mid-1970s.

How Many Different Cancers Are There?

There are over 100 different types of cancer that can be identified by examining cells from the tumour. Cancers are usually described by the part of the body from which they originated (liver, lung, etc.). However, as we shall see when we look at the molecular picture, that classification is beginning to be replaced by a genetic definition – that is, on the basis of specific mutations.

A further classification is based on the type of cell from which the tumour formed. The three main groups are as follows:

1. Carcinomas – cancers derived from epithelial cells. Skin is made of epithelial cells (epithelial cells are what you scrape off the inside of your cheek) but they also form the lining of all your organs – throat, intestines, blood vessels, etc. Cancers that arise in this type of epithelia are called adenocarcinomas. Carcinoma *in situ* is a pre-malignant change that happens in many cancers in which cells proliferate abnormally within their normal location: the epithelial cells show many malignant changes but have not invaded the underlying tissue. Ductal carcinoma *in situ* (DCIS) is one of the two most common forms of breast cancer, characterized by abnormal proliferation in the wall of the milk ducts. It carries a risk of developing into the invasive ductal carcinoma (IDC) form in which the cells are malignant. The majority of cancers (85 per cent) are carcinomas (e.g., breast, prostate, lung, colon).
2. Sarcomas involve connective tissue – that is, bone and soft tissues such as muscles, tendons and blood vessels. They are much rarer than carcinomas, accounting for less than 1 per cent of cancers, and are not thought to have a pre-malignant (*in situ*) phase.
3. Leukaemias (liquid cancers or blood cancers) and lymphomas are two groups of cancers arising in blood cells. Leukaemias affect bone marrow, whereas lymphomas arise in lymph node cells. The word 'leukaemia' comes from the Greek for 'white' (*leukos*) and 'blood' (*haima*). White blood cells are sometimes called leukocytes but, as that term covers all white cells, including lymphocytes, it's apt to be a bit confusing. Lymphomas are cancers of lymphocytes, the cells of the immune system that fall into the two main classes of T cells and B cells. The two major

divisions of this cancer are Hodgkin's disease (Hodgkin's lymphoma) and non-Hodgkin's lymphoma. This is our first meeting with our immune system, but we'll return to it as one of the hottest strands of current cancer therapy.

Is There A Difference Between Men and Women?

We're talking cancer here, of course, and it's pretty even-handed in its affliction of males and females, but men come off slightly worse (global new cases in 2018 were 9.5 million men vs 8.6 million women, and male vs female deaths were 5.4 million against 4.2 million). There's a general trend towards increased numbers of cancers that is more pronounced in females. However, survival rates are slightly higher in women.

Can You Catch Cancer from Someone Else?

The answer to this oft-asked question is: 'No, you can't.' But, as so often in cancer, the true picture requires a more detailed response – Alexander Pope might have approved – even though dealing with generalizations that aren't quite absolute tends to make scientists unpopular. It's not our fault! As Einstein more or less said, 'make it as simple as possible but no simpler'.

As we have noted, some human cancers arise from infection – notably by human immunodeficiency viruses (HIV) that can cause acquired immuno-deficiency syndrome (AIDS) and lead to cancer, and by HPV. We met HPVs just now and noted that, although they may have some rather unappealing effects – namely warts – most of the 100-plus HPV types are not life-threatening. However, and regrettably, 15 of them are. Of these the most important are types 16 and 18. These cause about 70 per cent of cervical cancers and a variety of other anogenital tumours, as well as some cancers of the mouth, voice box, windpipe and lung. The strains of HPV responsible for most cervical cancers are sexually transmitted – meaning that the causative agent (i.e., the *virus*) is transmitted, *not* tumour cells, and it is viral infection that can give rise to lesions that are the precursors of cervical cancer. We'll return to HPVs later, when we look at the molecular basis for their effects and how that has led to therapies that now offer the possibility of eliminating cervical cancer.

However, while humans cannot catch cancer from each other, there are three known examples in mammals of transmissible cancers in which tumour cells *are* spread between individuals: the facial tumours that afflict Tasmanian devils, a venereal tumour in dogs and a cancer passed between Syrian hamsters. Not to be outdone, the invertebrates have recently joined this select club and it has emerged that a variety of clams, mussels and other members of the bivalve mollusc family can transfer cancer cells between themselves. About a dozen species of these little beach-dwelling chaps have been shown to acquire a form of cancer through cells spreading from a single 'founder' throughout a population. This remarkable effect with, it is presumed, the inadvertent transport of infected molluscs on shipping vessels, has led to the spread of cancers from the Northern to the Southern Hemisphere and across the Atlantic Ocean.

Despite these rare events in the animal world, the key point to grasp is that humans cannot 'catch' cancer from each other in the way that, for example, influenza can be spread.

Do Metastases Metastasize?

Given that metastasis is the formation of secondary tumours from cells that have been shed from a primary tumour, an obvious question is 'Can cells similarly detach from a secondary and seed yet another site?' Mice at least can certainly carry out what Joan Massagué of the Memorial Sloan Kettering Cancer Center has called 'tumour self-seeding', whereby circulating tumour cells can colonize the tumours from which they originated. This can be tracked by tagging metastatic cells with a fluorescent label and inoculating them into mice, which reveals that extensive seeding by circulating tumour cells is a common occurrence.

This type of metastasis may occur in at least some human cancers and returning tumour cells may increase the aggressiveness of the tumour.

How Does Cancer Kill?

Given that tumours are abnormal growths of cells, one might expect them to have lethal potential if they interfere with normal function, rather as we noted for benign tumours if they compress adjacent tissues. This is a particular problem with brain tumours, the vast majority of which (about 90 per cent) are secondary cancers (i.e., metastases arising when tumour cells spread from another part of the body – commonly from breast or lung cancers). These secondary growths can create pressure on surrounding brain tissue, thereby affecting function. Bowel and gastrointestinal tract tumours, and also ovarian carcinomas, can obstruct the bowel, which would be fatal without surgical intervention. Damage to blood vessels caused by tumours can result in haemorrhage, particularly in the liver. Lung tumours can block lung function and some thyroid tumours can, in effect, cause strangulation.

Some liquid cancers can produce such an excess of white cells over red cells that the blood becomes very viscous and the supply of oxygen via the circulation is drastically impaired.

Human beings are, however, astonishingly resilient to organ damage: we can manage with half a kidney, and if we lose two-thirds of our liver it will regenerate itself. It is therefore relatively rare for cancers to kill via organ failure. Death usually results from secondary effects, principally infection by bacteria, such as *Escherichia coli* (*E. coli*) that can overwhelm the host even with antibiotic treatment. Cancer patients usually become more susceptible because the efficiency of their immune system declines.

About 50 per cent of cancer deaths are ultimately from malnutrition – a general condition of starvation and debilitation called cachexia (wasting syndrome) that develops in many other chronic diseases. In cancers, loss of appetite and inadequate digestion of food can occur. Cancer cachexia is poorly understood and there are no satisfactory treatments.

Can Plants Get Cancer?

Yes, even trees, corals and fungi can get a form of cancer. In plants and trees their abnormal growths tend to arise from infection by microorganisms (fungi, bacteria and viruses); insect infestation can also cause 'plant cancers'. They're not as serious as animal cancers because the cells of plants and trees have rigid cell walls and are locked into a matrix so they can't migrate. Thus, uncontrolled proliferation may result in the appearance of galls or burrs (burls), but not much damage is done.

Can We Cure Cancer?

No is the answer, notwithstanding the fact that, as we shall see, an increasing number of cancers can be treated with approaching 100 per cent success in terms of survival. The problem is that cancers are a fact of life – they're an inevitable result of how living organisms work – and in an increasingly elderly population, even if we have 'cures', in patients who are frail or in whom cancers are picked up at a very late stage, these 'cures' become either intolerable or ineffective. But in this book we will look at some of the ingenious methods being developed both to treat and to detect cancers. The expectation is that within the next 30 years the major cancers will be controllable and that it will become increasingly possible to 'nip cancers in the bud', so to speak – that is, to detect their presence ever earlier and target them before they become life-threatening. But we should always bear in mind that about half of all cancers are self-inflicted in that they arise from how we treat our own bodies.

Can Cancer be Modelled?

One of the longest running challenges in science has been to replicate cancer in the laboratory, a critical first requirement being to maintain animal cells under conditions that allow them to survive, grow and reproduce themselves. This is known as cell culture *in vitro*, first achieved in 1885 when the German zoologist Wilhelm Roux kept a piece of chicken embryo alive for several days by immersing it in saline (salt) solution. The first half of the twentieth century saw the development of *in-vitro* methods to the extent that by 1948 poliovirus could be grown in both human and monkey cells, a critical step in the production of the poliovirus vaccine.

In 1951 George and Margaret Gey derived the first human 'cell line' to be established in culture by taking cervical cancer tissue from Henrietta Lacks, a patient at Johns Hopkins Hospital in Baltimore. Named HeLa cells, these have become one of the best known and most widely used cell lines in biological research. A decade later, George Todaro and Howard Greene found that in cultures of mouse embryo cells most of the cells stopped growing after a short while, but some survived and grew rapidly. These cells could be removed when they completely covered the culture dish, then diluted and allowed to continue growing, a process that could be continued indefinitely. Todaro and Greene had shown that normal cells with a finite lifespan could give rise to 'established cell lines' that survived indefinitely.

The establishment of the basic principles for the *in-vitro* study of animal and plant cells has underpinned the acquisition of much of our knowledge of cell biology in general and of cancer biology in particular, although no *in-vitro* system has yet completely recapitulated *in-vivo* cancer. As we shall see, the twenty-first century has brought further progress in the shape of 3D cell culture technology and the capacity to do tissue engineering.

The use of mice is an important complement to *in-vitro* systems, either by studying the growth of transplanted tumour cells or by introducing predisposing mutations via genetic engineering. Although expensive and time-consuming, mice have provided considerable evidence about specific genes in cancer and are useful in testing anti-cancer agents.

The word 'model' may also refer to hypotheses or explanations – in this case as to how cancer cells originate and develop from normal cells in the process variously called carcinogenesis, oncogenesis or tumorigenesis. From this brief opening chapter it will already be evident that a basic model for cancer rests on the accumulation of mutations that ultimately perturb the normal processes of cell proliferation, cell death and control of cellular location. A vast amount of evidence supports this premise, although, as our story unfolds, it will become apparent that current knowledge is a long way from being able to explain all that happens in cancers. Inevitably, these shortcomings have led to alternative notions being put forwards – for example, that carcinogenesis arises from disruption of local tissue organization. As we shall see, it has indeed become clear that the tumour microenvironment – made up of a variety of host cells and tissues – is critical in the development of solid cancers. However, an overwhelming weight of evidence indicates that the fundamental driving force in the genesis of cancers is the acquisition of mutations in DNA.

Admiring the Picture

I promised that a clear picture would emerge from these answers, so let's just see what we've got! Cancer is the abnormal growth of cells, either in speed, location or both. It's caused by changes in our genetic material, DNA, collectively referred to as mutations. Mutations affect the activity of the proteins encoded by genes; for this reason, the term 'cancer gene' has come into common usage. It's a very handy expression but you need to bear in mind that there's no such thing. 'Cancer gene' is just a way of referring to a gene that has a perfectly normal, often essential, role in cellular control that has been changed either in the activity of the protein encoded or the amount of protein made, so that it can now help to drive cancer. Mutations occur randomly throughout life, from the point where sperm meets egg, and some infections may give a helping hand to this accumulation, as does smoking, one of the most common causes of cancer. Because cancers arise from mutations in our own DNA, they are not contagious: you can't pass 'cancer genes' to someone else, nor do they arrive from outer space.

Most cancers are diseases of old age, but they can occur at any age – so from conception to death we are engaged in a game of genetic roulette. By examining the affected tissues we can identify a huge number of different types of cancer. However, we will shortly come to the revolution that has enabled us to identify all the changes that have occurred in the DNA of an individual cancer, and we will have to grasp the astonishing fact that not only are *all* cancers different at the molecular level, but also that within a given tumour *each cell* has unique differences in its DNA. They're horribly varied and unstable – and hence very unpredictable. This quite staggering picture of evolutionary diversity has been acquired during the lifetime of the cancer within one person. It you wish, you could call it hyper-dynamic Darwinism (Figure 1.1).

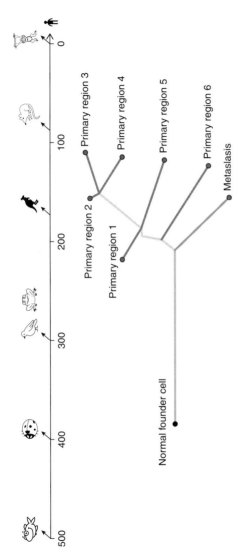

Figure 1.1 Species evolution versus cancer evolution. The top scheme is our family tree: the numbers are millions of years ago. The earth is about 4.5 billion years old, the first simple cells appeared around 3.8 billion years ago, and multicellular life started about 1,000 million years ago. The lower scheme represents the dissection of a tumour as an evolutionary tree. Six different regions of the same primary tumour and a secondary metastasis (M) are shown. The length of the lines between the branch points is proportional to the number of mutations picked up – thousands of them – on that stretch of the journey. (A black and white version of this figure will appear in some formats. For the colour version, please refer to the plate section.)

After that brief summary of the current position, it's timely to spend a moment (well, the next chapter) on a brief history of how we got here. In part it's an homage to all the patients, doctors and scientists who have gone before and struggled with this mightiest of problems. But it also enables us to see how mankind has applied his powers to making advances, however small or crude, and that gives an insight into how science works. That, in turn, is both useful and interesting because, like it or not, we live in a world increasingly dominated by science, and over the last 20 years cancer has become high-powered science in spades, so we need to be able to keep up!

2 Ancient History

You could plausibly claim that the story of cancer began in 1953. By that I don't, of course, mean that cancer only appeared as a disease in the middle of the twentieth century, but that it was not until the dawn of the age of molecular biology, heralded by the revelation of the shape of the DNA molecule, that we could approach the study of cancer at the molecular level. Even from that point it's been a mighty slow trek and we'll come in due course to the many highlights and some of the lowlights of that journey. Before that we should immerse ourselves in history for a while, partly because it's jolly interesting but also because we should pay homage to all those who went before us, battling with the disease either in themselves or as a major challenge to human ingenuity.

The earliest recorded account of cancer so far discovered is an Egyptian papyrus that was bought by an American called Edwin Smith on a visit to Luxor in 1862. Estimated to date from approximately 1600 BCE and written in the form of hieroglyphs, it was first translated in 1930 by James Breasted of the University of Chicago. It's a remarkable document, not least because it reads like a set of concise clinical reports revealing that, even at that time, rational, scientific practices based on observation and examination were in use. The record deals with 48 cases, one of which appears to have been an abscess that was hot to the touch and treated by cauterization using a red-hot iron. Another involved a 'bulging tumour on his breast' that felt very cool, with no sign of fever. For this, concluded the unknown physician, 'there is no treatment'.

In a remarkable complement to the Edwin Smith papyrus, in 2018 archae-ologists unearthed six bodies bearing signs of cancer in Egypt's Western Desert. These included a toddler with leukaemia and a mummified man in his fifties with bowel cancer. These intrepid anthropologists estimated that the lifetime cancer risk for those Egyptians was about 5 in 1,000, compared with 1 in 2 in modern Western societies – a difference in part due to their lower life expectancy of around 30 years. At about the same time, another group found two entombed mummies and applied the latest technology to show them to be a woman with breast cancer who died around 2000 BCE and a man with multiple myeloma who died around 1800 BCE.

All that, however, has been spectacularly upstaged by some palaeontolo-gists who, ferreting around in a field in what is now Germany, stumbled over the remains of a turtle who had been stricken with a kind of bone cancer common in humans today. Almost incomprehensibly, this reptile's tumour-bearing, fossilized femur turned out to be 240 million years old (Figure 2.1)!

The Greeks Had a Word for It

We can therefore be confident that cancer had been around for a long time before perhaps the most famous name in medical history made his appearance. The Greek physician Hippocrates, often called the *Father of Medicine*, lived around 400 BCE and is credited with being the first person to take a scientific view of disease as a natural process occurring within the body, as opposed to being inflicted upon us by some sort of magical, quasi-religious force. Hippocrates observed that tumours were often rich in blood vessels that, when the tumour was dissected, resem-bled the limbs of a crab and thus gave us the word 'carcinoma' from the Greek for crab 'carcinos'. The prefix 'onco' also comes from Greek, 'oncos' meaning a swelling. We'll meet shortly the phenomenon of controlled cell death, 'apoptosis', the Greek for the 'dropping off' of petals or leaves from plants or trees, that Hippocrates used to describe gangrene, in which tissues become black and decay as a result of infection or restricted blood supply.

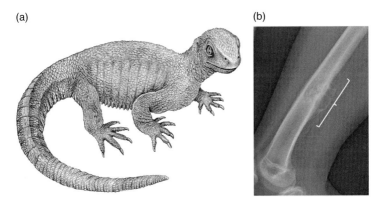

Figure 2.1   240 million years of cancer. (a) Now extinct, *Pappochelys rosinae* was a reptile closely related to turtles. A fossilized femur in a hind leg of one of these animals bearing an osteosarcoma, a highly malignant bone cancer, was excavated in 2013. (b) A recent image of a human osteosarcoma on the femur above the knee joint (the white bracket delineates the extent of the sarcoma). Although our mammalian ancestors diverged from reptiles in evolution about 220 million years ago, the two tumours look very similar. These bone cancers are relatively rare (about three cases per one million people worldwide every year). They tend to occur in teenagers and young adults and are usually found in the long bones (legs, but sometimes arms). Treatment may involve chemotherapy, surgery and radiation therapy.

Some 600 years later Galen, a fellow countryman of Hippocrates, emerged as a prolific writer and radical surgeon – he was the first to treat cataracts and to use the pulse as a diagnostic measure, and is also generally credited with being the first to use the word 'cancer', the Latin for 'crab'.

Following Galen, the Persian physician Avicenna (also known as Ibn Sina), one of the great scholars of the Islamic Golden Age, completed in 1025 the *Canon of Medicine*, which was translated into Latin to become the encyclopaedia of medicine in Europe throughout the eighteenth century. It includes the first description of surgery for cancer and also what may be the first recorded cancer drug treatment with an extract from the Middle Eastern plant *hindiba*.

Chinese Science

Remarkable though the efforts of these luminaries were, in thinking about ancient medicine the minds of most people would alight first on China, in part because we've gradually become aware that many scientific advances were made by the Chinese centuries before they occurred in Western countries. Much of our knowledge of Chinese science is due to the genius of one Joseph Needham, who first studied medicine but was then persuaded by Frederick Gowland Hopkins, head of the Department of Biochemistry in Cambridge, to focus his talents on research. Hopkins was to share a Nobel Prize for his discovery of vitamins (Nobel Prize winners whose work contributed to the story of cancer are listed after at the end of the References section at the end of the book). Needham made outstanding contributions to the fields of muscle biology and embryology, including writing the seminal three-volume *Chemical Embryology*. However, his life was turned upside down in 1936 when a Chinese graduate student, Lu Gwei-djen, arrived in his department. They fell in love and thus began Needham's obsession with China. By 1942, having learned to read and write Mandarin, Needham was travelling extensively in China as the director of the Sino-British Science Co-operation Office in Chongqing. Needham's extraordinary ability to absorb and distil information in every field from agriculture to medicine enabled him to write *Science and Civilisation in China*, an astonishing 24-volume compendium of the history of Chinese science.

In the field of medicine this includes a description of smallpox from the year 340 and of smallpox inoculation around the year 1570. This was by scraping off matter from smallpox pustules with a cotton swab that was pushed up the nose of the person to be inoculated. Recall that it would be 1721 before Dr Richard Mead tried his notorious reprise of this experiment on a prisoner in Newgate Prison, and 1796 when Edward Jenner showed that inoculating pus scraped from cowpox blisters conferred immunity. These steps were mankind's first venture into the field of immunology, of which much more later. However, Jenner's work very quickly made an impact on the figure of 10 per cent of Britons killed annually by smallpox at that time. Eventually, in 1980, the World Health Organization was able to certify the global eradication of this scourge. It's worth making the point again that eradication is not a realistic aim for cancer. The nature of these diseases dictates that our goal is control.

Not the least remarkable thing about *Science and Civilisation in China* is that Needham makes no mention of cancer, although there are references to what might be cancer in the *Huang ti nei ching* (The Yellow Emperor's manual of corporeal medicine, dated to 1500 BCE and thought to be the oldest medical textbook), and also in Sun-su miao's *Thousand Golden Prescriptions*, dating from the Tang dynasty (618–907). The earliest reference to herbal treatments appears to be from the Ming dynasty (1368–1644), by which time an extract from the flowering plant *Aconitum* (also known as aconite or wolf's bane) was in use as a cancer remedy. Chemical compounds in these plants have in recent times been shown to have inhibitory effects on the growth of cancer cells in the laboratory and remain in contention as potentially useful drugs. Other phytochemicals extracted on the basis of herbal medicine include curcumin, resveratrol, berberine and ginseng. Ginseng is a perennial plant containing phytoestrogens, plant counterparts of oestrogens, the growth-promoting hormones in mammals.

All of which reflects the fact that traditional Chinese medicine in the shape of herbal remedies is not only still going strong in China after 2000 years, but has also come into consideration in the West as potential sources of anti-cancer agents.

The Coming of Science

While Chinese science remained unknown in the west, the accessibility of the *Canon of Medicine* coincided with a surge of intellectual activity in the seventeenth century, most clearly marked by the founding of the Royal Society of London in 1660 under a charter granted by King Charles II. It was to be a forum for leading figures to exchange ideas about science in an informal way. Two of the first members were Christopher Wren, the great architect, and Robert Boyle, the founder of modern chemistry. The Royal Society published Isaac Newton's *Principia Mathematica*, and Newton himself was president from 1703 to 1727. Some 300 years later, in 2015, the society was to elect as president Venkatraman Ramakrishnan, a biological scientist who worked out the molecular detail of how proteins are put together – a topic we'll come back to shortly. Going back to the founders, one of the most outstanding was Robert Hooke, a polymath if ever there was one. Hooke worked with Boyle on a study of the physics of gases; as a skilled surveyor and architect he designed the Greenwich Royal Observatory and the Monument to the Great Fire of London; as an astronomer he explained that light behaves like a wave and that as things get hotter they expand. And, of course, he came up with a fundamental law of elasticity (how far a spring stretches depends on how hard you pull it). Moving into biology, he had a go at skin grafts and carried out the first recorded blood transfusion – on a dog.

However, Hooke's immense impact on biology came from starting to play with microscopes. Simple lenses for magnifying objects were in use by the late 1500s, and by 1624 another great polymath, Galileo Galilei, had devised a compound microscope (one with two lenses) – essentially the design used to this day. Galileo is of course best known as an astronomer, but Hooke's inclination was to look down rather than up, so to speak, and he produced the first images of things in the natural world that are too small to be seen by the unaided eye. Hooke's incredibly detailed drawings of what he saw include a flea, a louse and a gnat. These formed part of the first scientific best-seller, *Micrographia*, published by the Royal Society in 1665. In addition to various insects, Hooke pointed his microscope at plants, specifically at pieces of cork oak trees. It revealed regular structures that to Hooke resembled the shape of a monk's cell (Figure 2.2). Thus, for the first time was the basic unit of life revealed – and Hooke simply used the word 'cell' to describe it.

It was the best part of another 200 years before it emerged that cancer was a reflection of something going wrong with cells in animals. Nevertheless, Hooke's observations and the other great scientific events of that time opened the eyes of all with an interest in the natural world to the notion that, by observation and intelligent deduction, it might be possible to make sense of what hitherto had been explained only by religious mysticism.

Scientific Observation and the Practice of Medicine

Among the first to apply this approach to cancer was the Italian Bernardino Ramazzini, who noted in 1713 that nuns were particularly prone to breast cancer but rarely developed cervical cancer, and concluded that this might

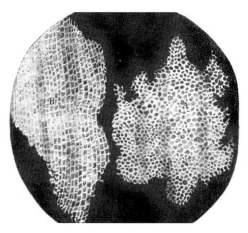

Figure 2.2 Hooke's drawing of cells.

have something to do with their lifestyle. Of course, his observations reflected the fact that cervical cancer most commonly results from the sexual transmission of a virus and that pregnancy is a strongly protective factor against breast cancer – and nuns are generally held to avoid sex and hence pregnancy.

Another astute scientific observer, the English surgeon Percivall Pott was one of the founders of orthopaedics and the first scientist to demonstrate that cancer can be caused by an environmental agent. He spotted that cancer of the scrotum was common in chimney sweeps. This was in 1775, when chimneys were swept by boys shinning up the vents, a job they often carried out naked. Those who were astute or coy enough to wear a leather garment about their nether regions were protected, and Pott, comparing the incidence of cancer with the modesty of the sweeps, concluded that soot accumulating in the folds of the scrotum caused cancer. This was the first identification of an occupational exposure to cancer-causing agents and it led to other such risks being recognized. In 1915 Katsusaburo Yamagiwa and Koichi Ichikawa at Tokyo University did the experiment of applying coal tar to the ears of rabbits, whereupon skin cancers developed. Soot and coal tar, like tobacco smoke, is a complex mixture and it took until 1930 before Ernest Kennaway isolated the first specific chemical – a 'carcinogen' – that can promote cancer. The carcinogen extracted from coal was a cyclic hydrocarbon (basically a ring of hydrogen and carbon atoms).

The Advance of Surgery

These were the first steps taken in unveiling the causes of cancer, but at around the same time other curious individuals were venturing into the cellular and surgical world of cancer. A notable early contribution came from the French physician René-Théophile-Hyacinthe Laënnec, who, in 1816, invented that most emblematic of medical instruments, the stethoscope. He first made a hollow paper cylinder and then a wooden tube to enable him to listen to the heartbeat of a young lady whom he described as having a 'great degree of fatness'. However, Laënnec's cancer connection was as the first person to describe melanoma, the abnormal growth of skin pigment cells, and to note that such growths could spread to the lungs. A few years earlier, in 1787, the Scottish surgeon John Hunter had carried out what is thought to be the first successful surgical removal of a melanoma. We should bear in mind that it was not until 1846 that William Morton, a Boston dentist, achieved a painless tooth extraction after administering ether to the patient. He went on to stage a demonstration of ether as a general anaesthetic at Massachusetts General Hospital.

The availability of general anaesthesia gradually, together with the slow acceptance of germ theory – that microorganisms can lead to disease – permitted increasingly radical surgery as a means of dealing with cancer. An early pioneer was Christian Albert Theodor Billroth, who, in 1874, carried out the first removal of the larynx for cancer and the first successful surgical treatment for stomach cancer. Billroth had also recognized that benign growths – polyps – on the lining of the bowel could turn into cancers, and he was the first to remove a bowel cancer. Billroth's first love was music, which he actually studied briefly at the University of Greifswald before becoming engrossed in medicine. In 1867 he became professor of surgery at the University of Vienna, where his musical gifts brought him into contact with Johannes Brahms. They became lifelong friends and Brahms rendered Billroth unique among cancer scientists by dedicating a string quartet to him.

It was perhaps fortunate that Billroth pre-deceased Brahms by three years, thus being spared the painful irony of seeing his friend die of liver cancer.

The outstanding American surgeon William Halsted became, in the 1880s, the leading exemplar of developing novel surgical procedures. He embraced the new thinking that cleanliness might be a good thing in surgery and came up with the idea of surgical gloves – it is said to protect the hands of his scrub nurse, who showed her gratitude by marrying him. Halsted is particularly remembered for the extreme methods he applied to breast cancer involving not only removal of the breast but also of much adjacent tissue. His teachings persisted into the early years of the twentieth century, when breast cancer surgery could extend to hip or shoulder amputation. In the twenty-first century this sounds seriously gruesome, but it was a perfectly logical response to what was known at the time. Halsted was aware that cancer lethality almost always resides in secondary tumours formed from cells that had seeded from the original primary growth. Breast tumours often spread to the lymph nodes of the armpit and to bone – a good reason for trying ever more extreme surgery to prevent recurrence. In his writing Halsted comes across not merely as an innovative surgeon, but also as a rigorous scientist and a concerned doctor. In discussing the difficulty of 'curing' breast cancer once it has spread, he demands 'incontrovertible proof' but concludes that 'even if the microscopic findings were confirmed by an able pathologist I should still feel that an error might have occurred, for example, in the labelling of the specimen'. Clearly a hands-on practitioner well aware that even in the best-run outfits mistakes can happen. We will come in due course to the tricky matter of cancer screening, but even 150 years ago Halsted understood the problems, asking:

> Shall we let women know that a dangerous process may be going on which they cannot detect, and keep them in a constant state of apprehension, or shall we encourage them to seek 'expert' advice which may be insufficiently expert, and expose them to the annoyance of repeated and useless examinations, each of which for only a brief period, if at all, would bring a measure of reassurance?

A further notable figure in the context of breast cancer is George Thomas Beatson, one of the leading clinicians at the end of the nineteenth century, who shortly after he graduated became house surgeon to Joseph Lord Lister. Beatson carried out the first operations to remove the ovaries as a way of treating women with advanced breast cancer. This approach evolved from his realization that one organ could exert control over another by means of chemical messengers released into the circulation. The messenger in question was the primary female sex hormone oestrogen, although it was to be 30 years before it was isolated. In a slightly different approach, Beatson treated breast cancers with extracts from the thyroid. Although he could not identify the agents he was dealing with, he was convinced that unravelling the chemistry – that is, the molecular processes involved – was essential in understanding cancer. He laid the foundations of hormone therapy, and it is wholly appropriate that he is remembered today in the names of two of the world's leading research institutes, the Beatson Institute for Cancer Research and the Beatson Oncology Centre in Glasgow.

Beatson's work on female cancers was complemented, albeit 50 years later, by that of the Canadian Charles Brenton Huggins. In 1941 he discovered that hormones could be used to control the spread of some cancers. The purification of oestrogen enabled Huggins to show that testosterone and oestrogens had opposing effects on prostate tumours, the implication being that oestrogens could block the growth-promoting effect of testosterone on these tumours and provide an effective treatment without the requirement for castration. For his efforts in launching the field of chemotherapy Huggins shared the 1966 Nobel Prize in Physiology or Medicine. We shall meet his co-recipient later in this story.

In parallel with these surgical advances, William Sampson Handley, described as 'the most fertile brain in British surgery', developed methods for killing cancer cells that remained after breast surgery by chemical treatment. He was doubtless aware that the Egyptians, somewhat hazardously, had used arsenic to destroy cancerous tissue. However, his method relied on the work of Sir Humphry Davy, who showed that acids reacted with metals to form salts and that zinc chloride had a strong caustic action on live tissues. The first clinical use of zinc chloride was in 1910, and it is used nowadays in a variety of commercially available pastes and salves for treating superficial skin cancers. Silver nitrate can also be used as a cauterizing agent and is a quick and effective treatment for mouth ulcers. However, it should be noted that neither of these chemicals is at all specific for cancer cells or ulcers: they kill normal tissue with equal efficiency.

Zinc chloride has two additional and very useful properties. It's a very good tissue preservative and, if you cut the tissue into thin sections, zinc chloride highlights the detailed structure of the constituent cells so that you can distinguish normal from tumour cells. This finding was the result of a chance observation by the American physician Frederic E. Mohs, and it led to the method of 'fixing' – slicing into thin sections and staining tissues – now used worldwide. He also applied these properties to devise what has become known as Mohs' micrographic surgery, the microscopic analysis of tumour samples that have been removed from the patient and stained to permit detection of cancer cells. The process of surgical removal, slicing, staining and microscopic analysis is repeated until no further cancer cells are found. For the most common type of skin cancer this method gives an almost 100% five-year cure rate.

The Coming of Cell Biology

You may have noticed that we have so far not mentioned the most significant scientific event of the nineteenth century, or of any other for that matter, the publication in 1859 of Charles Darwin's *On the Origin of Species*. Well you might say, wonderful chap though he was, his revolutionary concept doesn't have much to do with cancer, does it? Darwin's idea was that genetic evolution facilitated adaptation over time to produce organisms best suited to their environment. However, it took rather more than 100 years until we could analyse the genetic code carried by DNA, before Darwin's theory was finally transformed into the established basis by which all life has evolved. In a way that few would have expected, the revelation through DNA of the common ancestry of all living species has turned out to have a strong parallel in cancers. Extraordinarily, individual tumours expand from a single cell through random mutations in an effect we've already referred to as hyper-dynamic Darwinism – meaning that a tumour mass grows by accumulating mutations that help it survive in its individual niche. It's an effect closely resembling the way in which individual species adapt to their environment. Cancers therefore mirror their hosts in that both are products of the mindless process of natural selection.

It's only possible to perceive this by getting your hands on the DNA inside cells and determining the code (i.e., the sequence of bases), but, important though that is, the biological effectors are the cells themselves, and from Darwin's time onwards a galaxy of gifted individuals began to illuminate the biology of cells. By 1860, Rudolph Virchow, the German physician remembered as the *Father of Modern Pathology*, had concluded not only that 'all cells come from pre-existing cells', but that cancers derive from normal cells and that the disease was basically no more than the uncontrolled growth of cells. Virchow and the English doctor and scientist John Hughes Bennett had identified leukaemia as a blood disease. Karl Thiersch, who hailed from Munich and was the son-in-law of the redoubtable organic chemist Justus von Liebig, had shown that cancers metastasize through the spread of malignant cells, rather than through some unidentified 'juice' as Virchow had earlier proposed. Most notably, in 1869, Friedrich Miescher had isolated from the nuclei of white blood cells what he called 'nuclein' – something that eventually turned out to be DNA and associated proteins.

Louis Pasteur, together with Robert Koch (who isolated the bacteria responsible for tuberculosis, anthrax and cholera), showed that microorganisms caused many diseases, discoveries that eventually led to the introduction of antiseptic surgical techniques. Koch became a good friend of the German immunologist Paul Ehrlich, who developed antimicrobial agents to treat disease and, with Sahachirō Hata, discovered Salvarsan, the first effective treatment for syphilis.

Miescher's extraction of nuclein prompted closer examination of what we would now call the genetic material of cells – the individual chromosomes that together make up the store of DNA. Chromosomes were first visualized during the process of cell division (when a parent cell divides to form two new ones) in the middle of the nineteenth century, and they turn out to be surprisingly easy to see. All you need is a simple light microscope – à la Hooke – because, as cells divide, their chromosomes become highly compressed in preparation for being split between the two new cells. When these condensed chromosomes are spread on a glass slide, their gross features are clear, even at the comparatively low magnification of 100 times.

In 1890 David Paul von Hansemann described seeing chromosomes within human tumour cells dividing abnormally and came up with the term 'anaplasia', meaning 'to form backward'. This really means that the cells are not 'differentiated' – that is, they haven't developed into a specialized type of cell and they don't have the structural features of a normal cell. This really lies at the heart of cancer, a condition of cells ceasing to respond normally to the world around them and reproducing themselves without restraint.

Theodor Boveri, who came from Bamberg and studied the humanities before switching to biology, had already tracked the maturation of egg cells in worms and sea urchins, and he was not the last person to make decisive contributions to human cancer biology through seemingly humble organisms. He concluded that heredity is carried by chromosomes. Most of the leading geneticists of the time, including the American Thomas Hunt Morgan, didn't think too much of this idea. However, in the way that science often has of humbling people, it was Morgan himself in 1910, through experiments on flies very similar to the earlier work of Gregor Mendel in the 1860s on pea plants, who confirmed that genes do indeed reside on specific chromosomes. Mendel was the celebrated monk-turned-scientist, born in what is now the Czech Republic when it was part of the Austrian Empire. By tracking things like flower colours in the monastery garden at Brno, he was the first to show that inherited traits come in little packets – genes – and he worked out that genes come in pairs and are inherited as distinct units, one from each parent – albeit some 50 years before Johannsen coined the word. Funnily enough, despite his scepticism, Morgan is perhaps better remembered than Boveri because his name is used as the measure of distance along chromosomes (in practice, it's centimorgans).

Meanwhile, as Morgan was coming to the conclusion that Boveri might have been right after all, the man himself, building upon the work of Hansemann, pointed out in 1903 that a fault in the machinery controlling cell division might produce cells with an abnormal number of chromosomes or with chromosomes that had been damaged in some way – and such cells might in due course become cancerous. The essential idea was that a normal cell turns into a tumour cell when something goes wrong with its chromosomes. Boveri noted that there could be 'countless' different, abnormal chromosome combinations but that almost all of these would be fatal for the individual cell. In other words, only very rarely would a cell acquire a set of genetic changes that would allow it both to survive and to develop the properties that make a tumour. Since Boveri's time, microscopy has been transformed and you can now tag each chromosome pair with its own colour. Using the same tags to look at chromosomes from cancer cells provides stunningly beautiful confirmation of Boveri's insight. Imagine each normal chromosome as a sausage, each with its own colour: slice and shuffle, then glue the bits together to reconstruct chromosomes. The result is spectacular banding patterns in the tumour chromosomes caused by the shuffling of large chunks of DNA that goes on as almost all cancers develop (Figure 2.3).

Boveri's name will always be associated with that of the Kansas-born Walter Stanborough Sutton because together they established the basic picture of what happens when a cell is fertilized and starts to divide. Every normal cell has *two* sets of chromosomes: before it divides the chromosomes are duplicated and so each daughter cell ends up with a complete set. However, egg and sperm cells are formed so that each has only *one* set of chromosomes. Thus, when they fuse during fertilization, you get one cell with *two* complete sets of chromosomes. All of this provided the first clues that chromosomes were the carriers of heredity. Boveri also suggested that normal cells might

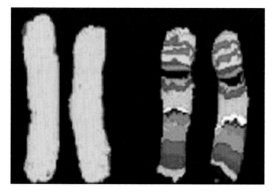

Figure 2.3 Chromosome rearrangements in cancer. Left: normal; right: tumour cell chromosomes. (A black and white version of this figure will appear in some formats. For the colour version, please refer to the plate section.)

have a built-in system that keeps them from dividing until they receive some overriding signal, a remarkably prescient thought some 60 years before we started to work out that a cell does indeed have to pass a series of 'checkpoints' before it can divide.

Around the beginning of the twentieth century, a new word began to edge its way into the scientific lexicon – biochemistry – referring to the study of what might grandly be called the chemistry of life, namely of proteins, sugars, fats and, of course, DNA – the molecules that make living systems. Biochemistry really began to emerge with the dissection of the major metabolic pathways that break down what we eat to provide energy and that build the molecules of which we are made. In the period following the First World War the pathway that breaks down the sugar glucose to release its energy, called glycolysis, was defined. The complete breakdown of glucose so that all its energy is released requires a second pathway, worked out by Hans Krebs in 1937 and often called the 'Krebs cycle'.

An indication of how important these molecular approaches to living systems were to be for cancer was not long in coming. By the 1920s, Otto Warburg had noticed that something odd happened to metabolism in cancer, and he showed that tumour cells get most of their energy from glucose (using the glycolytic pathway), despite the fact that it is less efficient than the Krebs cycle.

Warburg was part of an amazing scientific galaxy in the period from 1901 to 1940 in which one out of every three Nobel Prize winners in medicine and the natural sciences was Austrian or German. Born in Freiburg, he completed a PhD in chemistry at Berlin and then qualified in medicine at the University of Heidelberg. Fighting with the Prussian Horse Guards in the First World War, he won an Iron Cross and after the war he made so many contributions to biochemistry that he was nominated three times for the prize. He actually won the 1931 Nobel Prize for showing that cellular respiration, that is, oxygen consumption, involves proteins that contain iron. The 'Warburg effect', as the perturbation of metabolism in cancer became known, passed into near oblivion for the rest of the twentieth century, but in the last 20 years it has emerged into the sunlight again as metabolism has become a major focus of cancer research.

These individuals, through their technical genius and perception of how the human body works, established the basis of cancer both as a scientific discipline and as a field of clinical medicine. It remains true that the first method of treatment for many cancers is surgery, which is still the most effective method for enabling people to survive the disease. In a crude sense, therefore, 'cut, burn and hope' does sum up the foremost strategy. However, this aphorism undervalues the efforts of the people who laid the foundations of clinical methods that have become increasingly sophisticated. Surgeons now specialize in specific cancers using increasingly refined technical aids, and this trend towards ever greater, focused expertise will continue as novel methods for defining the geometry of tumours and also for removing them are developed. In parallel, the molecular science of cancer has established the field of chemotherapy that, together with surgery and radiotherapy, comprise the main tools of cancer treatment. It is the molecular biology that increasingly provides a rational basis for the development of drug treatments, tumour monitoring and tumour detection that we will focus on in this book.

3 Counting Cancer

Having looked at some of the basic facts, and indeed non-facts, of cancer it's tempting to go straight to the really exciting bits, namely the extraordinary molecular events that underpin the way animals work and the astonishing science that is gradually revealing what can go wrong to give rise to diseases in general and cancers in particular. However, before we indulge ourselves, perhaps we should pause for a moment to consider the question 'Why is cancer so important?' Well, you might answer, 'Because it kills a lot of us.' True indeed, but it transpires that it's well worth a bit of time and effort looking into that answer – and that means looking at a few facts and figures or, to make it sound even more off-putting, at the statistics of epidemiology. Before you run a mile on the grounds that maths makes your head ache, let me implore you to stay a while. I promise it will be worth a little pain, and the reason I'm so confident is that the facts of cancer, the sheer numbers, are so staggering, so mind-bogglingly overwhelming, that they begin to tell us something about the underlying causes of the disease.

The Big Picture

Let's start on the biggest stage by noting that the World Health Organization (more specifically the International Agency for Research on Cancer) estimated the global cancer burden in 2018 as 18.1 million new cases and 9.6 million deaths. That's a death every three seconds – and rising. By 2040 these numbers will morph into 29.5 million new cases and 16.4 million deaths. Worldwide, 1 in 5 men and 1 in 6 women develop cancer in their lifetimes and 1 in 8 men and 1 in 11 women die from the disease. About 44 million people are alive five years after a cancer diagnosis, a measure called the five-year prevalence. The terms '5 year survival' or '10 year survival' are the percentage of patients alive 5 or 10 years after diagnosis. However, caution is advisable when you view survival figures because they can vary greatly depending on, for example, the developmental stage a tumour has reached at diagnosis.

The top six cancer types in terms of incidence are lung, breast, bowel (colorectum), prostate, stomach, and liver (Figure 3.1). For the top three there were about two million diagnoses of each type in 2018. In the mortality league the order becomes lung, bowel, stomach, liver and breast, with prostate dropping to eighth. The shifts in relative positions of specific cancers between tables of incidence and mortality reflect causes and available treatments. For example, most liver cancers arise from infection by viruses – something that rarely happens in the developed world but is relatively common in Africa and in Asia, the latter contributing 72.5 per cent of global cases, with China having the largest number (almost 400,000) and Mongolia by far the highest age-standardized rate (ASR) (the ASR is the number of cancer cases per 100,000 of the population, allowing for the age distribution of that population). The disease remains essentially untreatable, hence its sixth position in the incidence league but its fourth as a global killer.

The Global Picture

The best way to look across the nations of the world is to use age standardization, a method that permits comparison of populations while making allowance for the different age profiles. In the age-standardized incidence league table, including all cancers in both sexes of all ages, Niger comes top (i.e., with the *lowest* rate in the world, of 78 new cases per 100,000 people). Australia props up the table with the *highest* rate at 452. The US rate is 362 and that for the UK is 320 (Table 3.1).

The corresponding age-standardized mortality rates reveal an almost complete change of nations at the top and bottom of the table, with Saudi Arabia (51 deaths per 100,000 people) in the top spot and Mongolia (176) at the other end. The UK (101) and the USA (86) are somewhere in the middle. Of

(a)

Estimated number of new cases in 2018, worldwide, all cancers, both sexes, all ages

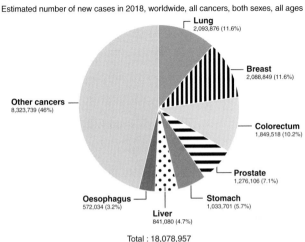

Total : 18,078,957

(b)

Estimated number of deaths in 2018, worldwide, all cancers, both sexes, all ages

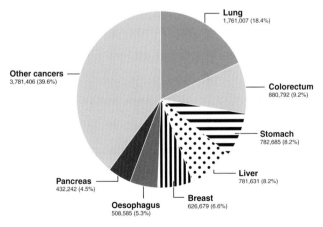

Total : 9,555,027

Figure 3.1 The 2018 global incidence (a) and mortality (b) for the top 10 major cancers. (A black and white version of this figure will appear in some formats. For the colour version, please refer to the plate section.)

Incidence		Mortality	
Lowest			
Republic of the Niger	78.4	Saudi Arabia	51.3
The Gambia	79.5	Nepal	54.8
Nepal	80.9	United Arab Emirates	55.9
Bhutan	81.9	The Congo	56.6
The Congo	84.4	Sri Lanka	57.2
Highest			
Denmark	351.1	Slovakia	141.3
USA	362.2	Montenegro	145.2
Ireland	372.8	Hungary	149.0
New Zealand	422.9	Serbia	151.7
Australia	452.4	Mongolia	176.2

Table 3.1   Cancer incidence and mortality worldwide 2020. Figures are estimated ASR for countries with the five lowest and five highest rates for all cancers, both sexes, all ages.

the European nations, Hungary (149) is the worst performer. The relative positions of countries in these two tables of incidence and mortality are one measure of how well each is doing in terms of cancer treatment, and in this context let's hear it for Australia with, as we've noted, the worst incidence rate but with a rate of 83 is elevated to 69th out of 185 in the mortality league.

As we've already made the point that the longer you live the more likely you are to get cancer, you would expect lifespan to be reflected in these figures – and indeed it is. For the countries mentioned, the average life expectancies are: Australia 83.3, UK 81.2, USA 78.9, Saudi Arabia 75, Hungary 77, Mongolia 70 and Niger 62.

You might also expect cancer rates to reflect the overall stage of development of a country, often assessed by the Human Development Index (HDI). Nations with high or very high HDI usually have 2–3 times the incidence rates of countries with low or medium HDI. The mortality rate differential between these two categories is generally smaller, in part because countries with lower HDI have a higher frequency of cancer types with poorer survival. In addition, poor development tends to be associated with less effective diagnosis and treatment.

The UK

The annual British contribution to these statistics is 363,000 new cases – nearly 1,000 a day – and around 165,000 deaths (450 every day). In line with the rest of the world, the big four are breast, prostate, lung and bowel cancer. All told, cancers account for more than one-quarter (28 per cent) of all deaths in the UK. For many years a little-publicized and rather disgraceful fact has been that UK cancer survival rates lagged somewhere around 10 per cent behind those for pretty much every other European country. In other words, for the last 50 years you were significantly more likely to die of cancer in the UK than in most European countries. In 1970 that 'significantly' was about 50 per cent. However, the most recent study has shown that rates have been falling since the early 1990s in both the European Union and the UK, but in the latter the drop is more pronounced (hooray!), albeit starting from a much higher rate (boo!). What's more, the pattern is similar for all cancers taken together. Thus, the gap has continued to narrow since 1970 such that now the UK appears to have well and truly joined the European Union in terms of overall cancer survival rates.

This good news is illustrated by the data for breast cancer shown in Figure 3.2, and before we go on we should note that, although breast cancer is the most common UK cancer, survival rates have doubled in the last 40 years such that the average 5-year survival rate for women with invasive breast cancer is now 90 per cent and the 10-year survival rate is 83 per cent.

But all is not well across the UK cancer scene. The National Cancer Intelligence Centre has produced a Cancer Atlas that compares incidence and death from the 21 most common cancers in 168 regions of the UK. For all cancers, the UK rates are 403 (incidence) and 175 (mortality), both per 100,000. The Atlas highlights regions of northern England and Scotland as cancer 'hot spots', illustrated by the corresponding figures for Glasgow (473 and 226). But the real shock comes when you compare the two areas at the extremes – Liverpool and Kensington and Chelsea. The comparative incidence rates are 498 and 296 and the death rates are 240 and 124, respectively. It's been pointed out that if you walk over Westminster Bridge from the House of Commons the life expectancy of the denizens around you drops five

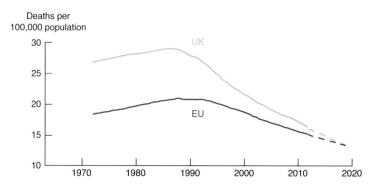

Figure 3.2 Breast cancer mortality rates for the UK and the European Union (age standardized). This shows the trend since 1970 and the predicted rates for 2019. The trends are similar for all cancer types taken together. (A black and white version of this figure will appear in some formats. For the colour version, please refer to the plate section.)

years. Go in the other direction to Liverpool and twice as many people are dying of cancer than in West London.

What's the Cause?

The stark finding is that where you live plays a huge part in how you fare against cancer. A combination of factors would appear to be responsible for this lamentable state of affairs. Clearly variations in levels of smoking, drinking, diet, and general social deprivation will play a role in cancer incidence, and they may also be reflected in later diagnosis. In addition, there are quite wide variations in the standard of treatment across the UK. It's a formal possibility that in the UK we suffer from types of cancer that are more aggressive (i.e., more difficult to treat) than occur in, for example, Europe, but there's no evidence for this. The problem is of our own making: how we look after ourselves and the efficiency of our health system.

The National Cancer Intelligence Centre estimates that if the worst UK areas were converted to the best, there would be 25,000 fewer new cases and 17,000 fewer deaths a year. A sensible start might be to concentrate cancer care into a smaller number of centres of expertise, along the lines proposed for heart disease.

As we turn our attention to the USA, we might note by way of consolation that eccentricities in cancer treatment across a nation are not exclusively a British problem.

The USA

The USA differs slightly from the global statistics in that while breast, lung, prostate and bowel are the most common types, these are followed by melanoma (skin cancer) in fifth place. The overall incidence rates for the USA in 2016 were 489 for men and 421 for women (per 100,000 of the population). However, this is complicated by the differing susceptibility of racial groups, of which white Americans are the majority (77%) with Hispanic and Latino Americans (18%) and African Americans (13%) making up the other major groups. White and black males are significantly more likely to get cancer than Asian-Pacific Islanders, American Indians, or Hispanics (i.e., white and black are above the average of 489/100,000). White women are also more susceptible, but black women are less so (421 vs 391/100,000). Mortality reflects incidence in that Asian-Pacific Islanders, American Indians or Hispanics have low rates. The rates for white males (193) and females (138) are close to the overall national averages, but for African Americans the male (234) and female (156) rates are significantly poorer.

Box 3.1 Lowest Cancer Rates by State

Bearing in mind the overall incidence rates of 489 for men and 421 for women (per 100,000 of the population), the states where cancer appears least often are New Mexico (M 390/F 366), Arizona (408/373), Nevada (409/382), Utah (440/376) and Hawaii (435/405).

Now, with the national death rates in mind (M 189/F 136), you might expect the same states to be at the top on the survival front – and indeed, by and large, they are: Utah (M 146/F 109), Hawaii (M 160/F 111), Arizona (165/122), Wyoming (164/123), New Mexico (167/123) and California (168/124). Wyoming and California just get ahead of Nevada. At the other end of the table? Kentucky with the worst incidence (558/479) and mortality (229/154) figures.

So, the message here is that if you're going to get cancer it's a good idea to be white and to go for one of the half-dozen or so with very good prognosis rates. The top ones are (with the five-year survival rate in brackets for white Americans and African Americans): prostate (100/98%), thyroid (98/97%), skin melanoma (93/77%), breast (90/79%) and Hodgkin lymphoma (89/83%).

The figures provided by the American Cancer Society reveal the British problem we've already highlighted: where you live is important. In summary, you're least likely to get cancer in New Mexico, least likely to die from it in Utah and most likely to get it and die from it in Kentucky (Table 3.2).

This American picture is consistent with the causes we described for the variation in rates across the UK. For all cancers, the UK incidence rate (320 per 100,000) puts the UK with the lowest USA states, and the mortality figure (101) puts the UK just outside the top group – that is, a bit worse than California. The Royal Borough of Kensington and Chelsea remains top for survival – even better than Utah! As for Liverpool – maybe it should be twinned with Jackson, Mississippi.

America's well-publicized black–white gap in cancer survival (and it's a feature of life expectancy in general) is a problem linked, of course, to the wretched socio-economic issue. It's notable that 4 states among the 10 highest in terms of black population (Mississippi, Louisiana, Alabama, and Delaware) are all at the wrong end of the death rate table. But, as with prosperity, anomalies appear when you try to make these simplistic associations. Thus, Maine (seventh lowest black population) is in the worst groups both for incidence and mortality. Connecticut is unique in being the only state in the highest incidence and lowest deaths groups. Do you think that might have something to do with it ranking alongside California in the wealth league?

Counting the Cost of Cancer

One cannot begin to grasp the cost that lies behind the accounting of this ongoing tragedy in terms of the suffering and misery involved, and it would be understandable to ask if we're getting anywhere at all with cancer. After all, it was in 1971, nearly 50 years ago, that President Nixon famously committed the intellectual and technological might of the USA to a 'War on Cancer',

Incidence

	Both sexes		Males		Females	
	Highest	**Lowest**	**Highest**	**Lowest**	**Highest**	**Lowest**
	Kentucky 510	New Mexico 361	Kentucky 558	New Mexico 374	Kentucky 479	New Mexico 355
	West Virginia 482	Arizona 375	Louisiana 549	Arizona 402	West Virginia 458	Arizona 355
	New Jersey 482	California 392	Mississippi 540	Colorado 406	Maine 454	Utah 365
	New York 481	Colorado 394	Arkansas 538	California 416	Pennsylvania 454	Texas 369
	Louisiana 479	Utah 397	New Jersey 530	Nevada 422	New York 454	Wyoming 377

Mortality

	Both sexes		Males		Females	
	Highest	**Lowest**	**Highest**	**Lowest**	**Highest**	**Lowest**
	Kentucky 186	Utah 121	Kentucky 229	Utah 140	Kentucky 154	Hawaii 106
	Mississippi 183	Hawaii 129	Mississippi 225	Colorado 153	West Virginia 154	Utah 106
	West Virginia 179	Colorado 131	Tennessee 215	Hawaii 158	Oklahoma 151	Arizona 115
	Oklahoma 177	Arizona 135	Oklahoma 211	Wyoming 158	Mississippi 151	Colorado 115
	Louisiana 175	Wyoming 136	West Virginia 211	Arizona 159	Louisiana 149	Wyoming 116

Table 3.2 Cancer rates in the USA. The five states with the highest and lowest age-adjusted rates of new cases and deaths are listed. Figures include all races and ethnicities and all types of cancer. Figures are for 2017, the latest year for which data are available from the Centers for Disease Control and Prevention.

saying, in effect, 'Let's give the men in white coats pots of money to sort it out pronto.' There's no question that large amounts of money have been forthcoming. In the USA, the National Cancer Institute has spent some $90 billion on research and treatment during that time. The annual expenditure on cancer research is over $5 billion, and the cost of cancer care in the USA is about $150 billion each year. In the UK, Cancer Research UK committed £546 million to cancer research in the last year alone. In addition to research funding, just the four main cancers cost the National Health Service about £1.5 billion every year.

Despite all of this there were 2,129,118 new cancer cases and 616,714 cancer deaths in the USA in 2018. Heart disease claimed about 635,000 people, and no other cause approaches these two. In proportion to the population, the UK picture is very similar with, for example, 164,000 cancer deaths in 2019.

In the face of all these monstrous figures you might be forgiven for wondering if we've made any progress at all, and you might also be asking yourself if cancer has always been such a scourge of mankind. After all, the example of the Teutonic turtle tells us that cancer has been around in animals since before the age of the dinosaurs – and indeed we know that they too were sufferers (there's evidence from spinal column X-rays that at least some members of the dinosaur family were prone to cancer – the hadrosaurs or 'duck-billed dinosaurs' seem to have been particularly unfortunate in this respect).

At the beginning of the twentieth century a review of the rather sketchy cancer statistics then available noted that, although there had been increases in the cancer death rates in a number of American cities and in England and other European countries, cancer mortality did not appear to be rising dramatically. In fact, the author, Wendell Strong, felt sufficiently confident to entitle his 1922 study 'Is Cancer Mortality Increasing?' He may have been somewhat uneasy, however, for he concluded: 'Lest what I have said be misinterpreted I would add that such a conclusion does not lessen at all the seriousness of the cancer problem. It merely holds out hope that the terrible scourge will not increase without limit.'

Well, there's nothing like optimism, and Strong's hope was nothing like the reality that unfolded. The lifetime risk of cancer increased from 39 per cent for men born in 1930 to 54 per cent for those appearing in 1960. For women the increase was from 37 per cent to 48 per cent. Over the period from 1935 to 2010 the age-standardized mortality rates for men went from 210 to 450 (per 100,000) and for women from 160 to 350 in more or less steady progressions. We'll look shortly at possible causes, but perhaps the most obvious is overall life expectancy. In the Bronze Age the average lifespan was 18 years: the average worldwide is now 66 – up from about 35 in the early twentieth century. So, of course, the cancer and heart disease figures are rising and the grim prediction for 2040 (in The Big Picture) is perhaps not so surprising. In the UK, lifespan went from 45 in 1900 to 81 years in 2019. The corresponding US figures are 49 and 79. As we've already noted in Chapter 1, that's more or less doubling the time available to pick up DNA damage.

Have We Made Any Progress?

In the USA the past 15 years have seen cancer incidence decline in men by about 2 per cent a year and stabilize in women. The death rate (meaning the number of deaths due to cancer in the population during one year) has dropped by a little under 2 per cent for both sexes. Much of this reduction is down to the decrease in lung cancer deaths, in turn linked to the long-term fall in smoking. That may not sound like a huge effect, but it translates to over two and a half million fewer cancer deaths between 1991 and 2016 than would have been expected if death rates had remained at their peak. In the UK, mortality rates for all cancers combined are projected to fall by 15 per cent between 2014 and 2035.

We noted that cancer death rates have historically been higher for African Americans than for white Americans, but over the past few decades that gap has been narrowing. In particular, the rates for three major cancers (lung, prostate and bowel) are dropping faster in African Americans. Notwithstanding that, there has been a fall in US life expectancy in each of the last three years. This is presumed to be a consequence of what have been called 'diseases of despair' – drug overdose, alcohol-related liver failure and suicide. In these pages we will, of course, talk a lot about exciting advances in treatment, but it's important to note that, so far, the greatest impact on cancer has come from targeting some of the things that are within our control, most notably smoking.

It is inevitable that reviewing cancer tends to focus on the developed nations – after all, that's where the money is to fight the 'war'. However, we should end this review of numbers by going back to where we started. The frightening trends shown in the global picture have, of course, many causes, including population growth and increasing life expectancy. But it really should focus our attention to reflect that if we could ban the use of tobacco and reduce the consumption of red and processed meat worldwide, we would cut new cancers by more than a half – a far more dramatic effect than has been achieved thus far by all the wonders of science.

In addition, it is within our powers to do something about the contribution from obesity – the biggest preventable cause of cancer after smoking, heart disease, high blood pressure and pollution. If we could only give the world clean drinking water we'd cut another 20 per cent of cancers – and incidentally put a stop to the deaths every year of 1.5 million children from diarrhoea, largely caused by picking up bugs through living in insanitary conditions.

It's already clear from the epidemiology that social and economic development is having an impact on certain types of cancer. In particular, as economies grow, cancers caused by poverty, including those due to infections, are being supplanted by those associated with the lifestyles of industrialized countries. The marked decline in lung cancer that has come from restricting smoking in North America and Northern Europe is heartening, but will be offset in years to come by the increase in tobacco-related cancer in China. Lung cancer is the most commonly diagnosed cancer, and the leading cause of cancer mortality in China. In 2018 there were 690,000 deaths, but if current trends continue, China's annual death toll from tobacco will reach two million by 2030 and three million by 2050.

More generally, Asia, with nearly 60 per cent of the global population, will continue to dominate the statistics of cancer. Currently 48 per cent of new cases and 57 per cent of deaths are in Asia. The corresponding figures for Europe are 23 per cent and 20 per cent, respectively, although only 9 per cent of the global population is in Europe. This evidence of the effect of the modern way of life is perturbing to say the least. Finally, in Africa as in

Asia, the proportions of cancer deaths are higher than the proportions of new cases. This disparity is due to there being a higher frequency of certain cancer types associated with poor prognosis and higher mortality rates in these regions, as well as the more limited screening and diagnosis facilities together with less effective treatment. Limited resources will come under increasing strain, not least because of the effects of cigarette smoking that in Africa will be an even greater long-term problem than in China.

Not a Pretty Picture

You couldn't describe the numerical story of cancer as beautiful – it's just scary – but one thing it does bring home is the importance of our subject. Cancer will make a significant impact on the lives of practically every human being on our planet. The main message from the numbers is that, wherever you live, the dominant forms are lung, breast, bowel, prostate, stomach and liver cancers. The pattern of cancer types is fairly similar across the developed world, but shows significant differences across the globe. This tells us two key things. First, cancer development is underpinned by similar, cumulative molecular events that arise inevitably as a consequence of animal biology and, as human beings are about 99.9 per cent identical in their DNA, we're all at risk. Second, our lifestyle – where we live, what we eat, etc. – plays a big role in cancer. In other words, to a considerable extent, cancers are self-inflicted.

Advances in treatment mean that for about half a dozen major cancers in the developed world the five-year survival rates are now over 90 per cent. Despite this wonderful progress, the fact remains that we have it in our own hands to cut cancer rates by over 50 per cent by acting on the knowledge we have of what broadly might be called lifestyle factors.

The pursuit of a cure for cancer has been a uniquely powerful driver in revealing how human beings work. It's fortunate that in telling the story of cancer we are looking at life itself because, as we shall see, there is no aspect of the molecular clockwork of life that isn't involved in these diseases. It is endlessly complex, ever compelling, often beautiful and always fascinating and, as answers have emerged to the puzzles of cancer, so too have many of the secrets of life been unveiled.

4 From DNA to Protein

In the opening 'question and answer' chapter we blithely asserted that most folk know that genetic material is made of DNA – the stuff of heredity – and that it is damage to DNA (mutations) that cause cancer. As we noted, these gigantic molecules are made up of huge numbers of four small chemical units (the bases A, C, G and T), linked together in two chains. In humans there are about 3,000 million bases in each chain. It is the sequence of these letters in DNA that forms a code telling the cellular machinery which proteins to make – proteins being the things that do all the work and hence make individual cells and animals what they are.

Given that cancers are driven by changes in DNA, the completion of the first sequence of the human genome in 2003 was a critical event in cancer research. This has been followed by incredible technical innovations that have permitted thousands upon thousands of human genomes to be sequenced. A major aim is to compare normal with tumour sequences, from which databases for all the major cancers are being compiled to identify mutations and hence possible drug targets for treatment. So important have these advances been, both for cancer and for our understanding of all biology, that it's worth sketching the sequencing pathway that has led to where we are now. It's a route that has seen expressions like 'it's in our DNA' and, thanks to COVID-19, 'PCR tests' become part of our normal vocabulary.

Atoms and Molecules

To begin we might remind ourselves that an atom is the smallest constituent unit of ordinary matter. A chemical element is a substance consisting of only one type of atom – there's currently 118 of them to be found in Dmitri Mendeleev's periodic table. Molecules are formed when two or more atoms are joined together by covalent bonds that stabilize the structure through the sharing of electrons. Compounds are chemical substances made up of two or more different chemical elements. A molecule of water is two hydrogen atoms joined to one oxygen atom by two covalent bonds – so it's H_2O. In Robert Schoenfeld's wonderful little book *The Chemist's English* he thinks of atoms as words and molecules as sentences – and along the way comes up with a brew of chemistry and English that's both funny and informative.

Our Genetic Material

Fertilization is the fusion of egg and sperm cells to merge two sets of chromosomes, each made of a single molecule of DNA plus some proteins. In humans this gives rise to 46 chromosomes, 22 (called autosomes) in which one half of the pair has been inherited from our mother and the other half from our father. In addition, we have two sex chromosomes (X and Y in men, two Xs in women), with Y being inherited from our father. Because chromosomes come in pairs (apart from X and Y), we normally have two copies of each gene (unit of heredity), called alleles (from the Greek *allelos*, meaning 'each other').

DNA is a nucleic acid (named from its discovery in the nucleus). The building blocks for DNA are bases linked to a sugar and a phosphate group (called nucleotides). A phosphate group is made up of one phosphorus and four oxygen atoms. Sugars are sweet-tasting carbohydrates with the general formula $C_n(H_2O)_n$. Table sugar is sucrose (fructose + glucose).

DNA was shown to be the material that chromosomes and genes are made from by the Canadian Oswald Avery and co-workers in 1944. In 1952 Alfred Hershey and Martha Chase confirmed DNA to be the carrier of genetic information by showing that viruses that infect bacteria (bacteriophages) do so by injecting their DNA into the recipient cell. Hershey went on to share a Nobel Prize with Max Delbrück and Salvador Luria for their work on the structure of viruses. Avery didn't get a Nobel Prize – but he does have a crater on the moon named in his honour.

How DNA is put together is, of course, important but the only thing that really matters is that the sequence of bases carries the genetic information that instructs cells to make proteins by gluing small building blocks (amino acids) together into huge chains. Everything does indeed stem from DNA, although it makes up only 0.25 per cent of the mass of a cell. How does that work?

The Double Helix

It's widely known that DNA in cells forms a double helix, and almost as well known that two chaps, James Watson and Francis Crick, worked this out in Cambridge in 1953. They built their celebrated model using X-ray crystallography data acquired by Rosalind Franklin and Maurice Wilkins at King's College London. For the structure to be stable, Watson and Crick concluded that the bases in one chain poke into the middle of the double helix (a little bit like the treads on a spiral staircase) and these pair, in a weak interaction, with bases on the opposite chain (Figure 4.1). The weak interaction is a hydrogen bond. Much weaker than covalent bonds, hydrogen bonds create the liquid state of water because the hydrogen atoms of one water molecule are attracted to the oxygen atom of a nearby molecule. Although individually weak, numerous hydrogen bonds between the base-pairs of two DNA chains makes the double-stranded form very stable. Stable, but not unbreakable – just warm up double-stranded DNA and the two chains fall apart. Crick's romantic analogy was between the two chains and embracing lovers – separable because the bonds forming the individuals are stronger than those between them.

An absolutely critical point is that bases do not pair up randomly: A only pairs with T, and C only pairs with G. Thus, the order of bases in one chain (that is, the sequence) determines the order in the other chain. For example, if there's a sequence TAGC in one chain, the weakly paired bases in the other will be ATCG – a complementary sequence.

Watson and Crick published their little paper in the journal *Nature* in 1953. A couple of years later Francis Crick gave a memorable talk to the *RNA Tie*

Figure 4.1   Sketch of DNA made by Odile Crick, Francis Crick's wife, for Watson and Crick's 1953 paper. The two ribbons represent the phosphate sugar backbones of the intertwined helices and the horizontal rods are the pairs of bases holding the chains together (A and T, C and G). The arrows show that the two chains run in opposite directions. The 'major and minor grooves' that arise from the orientation of the base-pairs across the helix are evident. These grooves separate the two sugar–phosphate backbones from each other, thus exposing bases for interactions with proteins.

Club in Cambridge, in which he outlined how the structure of DNA offered mechanisms by which essentially all the fundamental processes of life might work. If the double helix was 'unzipped' each strand could provide a template for a new strand to be made with a complementary sequence of base units. When cells divided, two identical sets of double-stranded DNA would then be available, one for each new daughter cell. Pulling the two strands apart would also make them available to be copied into an intermediate that could carry genetic information to ribosomes, the cellular machines that make proteins. The intermediate turned out to be ribonucleic acid (RNA), structurally similar to DNA but with a fifth nucleotide (uracil, represented by U) substituted for thymine (T) and containing the sugar ribose. The sugar in DNA is deoxyribose ribose, which is ribose lacking one oxygen atom.

Crick had the idea that a set of adapter molecules must exist that could pick up amino acids and present them to the ribosome in a way that enabled the

(a)   (b)

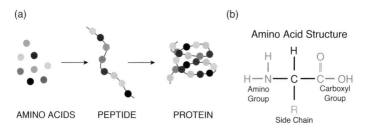

Figure 4.2   (a) Strings of amino acids make peptides and proteins. (b) Amino acids are molecules made up of an amine (–NH$_2$) and a carboxyl (–COOH) group attached to a carbon atom that is also linked to a side-chain (R group) specific to each of 20 different amino acids. (A black and white version of this figure will appear in some formats. For the colour version, please refer to the plate section.)

correct one out of the 20 or so in the cell to be joined to the growing protein (Figure 4.2), as specified by the DNA/RNA sequence of a gene.

All of this duly came to pass in a spectacular series of discoveries that launched the age of molecular biology. The first, in 1958, was the experiment in which Matthew Meselson and Franklin Stahl showed that each of the two double-stranded DNA helices arising from DNA replication comprised one strand from the original helix and one newly synthesized – 'semi-conservative replication'. John Cairns, the British physician and molecular biologist, described this as 'The most beautiful experiment in biology', as vividly recorded by Horace Judson in his wonderful account of those days, *The Eighth Day of Creation*.

Deciphering the Code

In 1961 Sydney Brenner, François Jacob and Matthew Meselson discovered messenger RNA, the intermediate nucleic acid predicted by Crick that carries the code of genes to the ribosome. Fifty years later Venkatraman Ramakrishnan and his colleagues resolved the structure of ribosomes and how their proteins and RNAs link amino acids together in the order specified by messenger RNA (mRNA) during protein synthesis (translation). Ribosomes move along mRNA molecules sequentially, capturing transfer RNAs carrying the appropriate amino acids to be matched by base-pairing through the anti-codons of the tRNA with successive triplet codons in mRNA. Simultaneously came the revelation that the codons used by the genetic code comprise three nucleotide bases, each codon corresponding to an individual amino acid. Marshall Nirenberg and Heinrich Matthaei deciphered the first of the 64 triplet codons, leading to the determination of the complete genetic code by which every amino acid is encoded by sequences of three bases in RNA. RNA is made by the action of an enzyme called RNA polymerase, an assembly of proteins that transposes the DNA code into a complementary version in RNA in a process called transcription. Proteins are then made by ribosomes reading the triplets of mRNAs – a step known as translation.

Despite the importance of DNA, it's proteins that do the work, enable us to function and make us what we are. Proteins make up 18 per cent of us – most of the rest (70 per cent) is water – and are made of long chains of amino acids held together by covalent (peptide) bonds, so they're 'polymers'. Covalent bonds are about 100 times stronger than any of the other forces that hold atoms together. Because these bonds are so strong, bond-making and bond-breaking in biology is given a helping hand by enzymes – almost always proteins – that significantly speed up the rate of chemical reactions in cells, enzymes being catalysts.

The Central Dogma

Thus were revealed the pieces of molecular Lego upon which all life depends, encapsulated in what has become known as the central dogma of molecular biology: *DNA makes RNA makes protein*. It had all occurred in an amazingly short space of time after the resolution of the structure of DNA. The story goes that when Watson and Crick finally convinced themselves that they had worked out DNA they retired to *The Eagle* pub for lunch, where Crick announced to the startled clientele that they had just discovered the Secret of Life! Francis Crick, notoriously, is not remembered for either his reticence or modesty, and his outburst must have been met with shaking heads and shrugged shoulders from the assembled lunchers. However, with the benefit of hindsight, it's hard to argue with his assessment.

Coding Power

All proteins are made from essentially 20 different building blocks (amino acids). The four bases in DNA and RNA, read in groups of three, can generate $4 \times 4 \times 4 = 64$ different triplet 'codes'. This means that, for just 20 amino acids, the genetic code has spare capacity – which leads to redundancy whereby some amino acids have more than one triplet codon. It also provides for some 'punctuation' – 'stop' signals indicating the end of a stretch of a coding sequence.

This simple sum tells us that just four letters is quite enough to tell the machinery that makes proteins what to do next – that is, which of the 20 units to tack on to the growing chain. The only other question is equally easily answered – albeit that the answer taxes the limits of the imagination. How many proteins can cells make? The smallest protein has three amino acids (called glutathione, with the sequence cysteine, glycine and glutamate), but in making proteins a cell can specify *any* of the 20 amino acids in *any* order and repetition is allowed. Hence 20^3 or $20 \times 20 \times 20 = 8,000$ *different* mini-proteins of three amino acids could be made.

After glutathione the next smallest protein (peptide) made by humans is oxytocin. It's a hormone of nine amino acids released into the bloodstream during labour. It also stimulates lactation after childbirth so mother can bond with junior, keeping them off the bottle – at least until they're a teenager. But the thing here is that the number of possible variants of a nonapeptide is 20^9 – that is, 20 multiplied by itself nine times: 512,000,000,000. But that's fine because your oxytocin gene encodes exactly the right sequence of amino acids out of the 512 billion possibilities for nine amino acids.

All of which makes the point that sequence determines function. The first person to work out the sequence of a protein was the extraordinary British biochemist Fred Sanger. Sanger chose insulin to sequence because, at 51 amino acids, it's one of the smallest proteins. Applying the rule of raising 20 to the power of the number of amino acids, you could have rather more than 2 followed by 66 zeros different proteins of that size! Sanger's painstaking work revealed the precise sequence of amino acids in insulin that enables it to regulate how the body uses and stores glucose and fat.

The critical point that emerges from this is that, in practical terms, the four-letter code of DNA that instructs the gluing together of just 20 different amino acids can make an infinite number of different proteins.

Shape Is All

Having an infinite palette of proteins is handy if you're going to create the entire living world, but how do they do it? The key lies in the fact that each of the 20 amino acids has slightly different chemical properties. These affect how the flexible protein chain will fold up into its final 3D shape, and it is this shape that determines what the protein does. Many proteins are more or less spherical, or globular, but in fact they come in all shapes and sizes – cables, bridges, etc. – and work through local regions on their surface that can interact with other molecules. With that picture in mind, it's easy to see that even changing one amino acid (by acquiring a mutation in DNA) might have drastic effects on activity – while, on the other hand, it might have no effect at all. The application of X-ray crystallography to proteins has given an unprecedented insight into how they work. Much of that story involves Max Perutz, one of the most influential biological scientists of the twentieth century. His work revealed the fantastic flexibility of the protein haemoglobin that enables it to pick up oxygen very efficiently in the lungs and to release it equally efficiently in tissues where the oxygen level is much lower. He also showed that the relatively common disease sickle-cell anaemia arises from a mutation of just 1 in 4,000 bases of the haemoglobin gene that changes a single amino acid. Many examples of single mutations that promote cancer have since emerged, as we shall see.

The legendary Perutz christened proteins 'the machines of life' back in the 1940s. His lifetime's work in revealing the secrets of proteins as flexible, adaptable, breathing structures has given us a background against which to think about how, from a potentially infinite pool, there has emerged proteins so stunningly precise and so versatile that they can make manifest the entire living world.

Controlling RNA Expression

A gene is usually defined as a sequence of nucleotides in DNA that can be transcribed into a complementary RNA sequence and subsequently translated into the amino acid sequence of a protein. A remarkable feature of genes is that, instead of being single blocks of a continuous coding sequence, they are interrupted by stretches of non-coding sequences. The non-coding bits are called introns. The regions of coding sequence are exons (from 'expressed region'), and it is these that are translated into proteins. So, coding DNA actually comes in fragments scattered along chromosomes. When a gene is 'switched on' (expressed), the entire stretch of DNA (exons and introns) is copied to make an RNA version called the primary transcript. Introns are then cut out and the exons stitched together to make the continuous sequence of triplet codons that is messenger RNA. The evolutionary advantage of this complexity is that, by varying the stitching pattern, more than one type of protein can be made from a given gene.

Primary control of this critical process of transcribing DNA is in the hands of 'promoters', regions a few hundred base-pairs (bp) long, just before the transcription start sites of genes. Specific sequences in promoters provide sites of attachment for proteins called transcription factors that determine whether a gene is transcribed or not. Additional control regions called 'enhancers' are short (50–1,500 bp) regions of DNA that can be bound by a subset of transcription factors called 'activators' to increase transcription of a particular gene. Conversely, transcription factors that bind to regions of DNA called 'silencers' to reduce transcription are called 'repressors'. In short, the production of RNAs from which proteins are synthesized is precisely regulated by the availability of transcription factor proteins that interact with DNA.

The Road to Sequencing DNA

The story so far has outlined how the basic processes of transcription and translation permit information flow from genes to proteins and noted that mutations in DNA can produce abnormal proteins with perturbed functions that could contribute to cancer development. The Human Genome Project and the ensuing sequencing revolution has led us into a new era wherein, for the first time, we can begin to grasp the true complexity of cancer and why it has been such an enigma.

We've encountered the outstanding South African Sydney Brenner playing a key role in working out the genetic code, but the question that preoccupied him was 'How do you get from DNA to a complete animal?' In other words, can you track where every cell comes from – how different types of cells (cell lineages) emerge from a single fertilized egg to make the complete animal? Brenner had decided that the ideal 'model organism' for this job was the roundworm *Caenorhabditis elegans*. This tiny creature, just 1 mm long and soil-dwelling like any self-respecting worm, was appealing because it is transparent and most adult worms are made up of precisely 959 cells. Simple it may be, but this worm has all the bits required to live, feed and reproduce (i.e., a gut, a nervous system, gonads, intestine, etc.). For this task Brenner recruited John Sulston, who had studied in Cambridge and then worked at the Salk Institute in La Jolla, California, to Perutz's institute in Cambridge. Their incredibly painstaking efforts resulted in the mapping, from fertilized egg to mature animal, of how one cell becomes two, two becomes four and so on to complete the first 'cell lineage tree' of a multicellular organism. One of the most startling findings was that during development 131 cells (precisely) disappear from the embryo as a result of 'apoptosis' – the controlled programmed cell death that we met in Chapter 2. This was the first revelation that, counterintuitive though it may be, cell death is part of normal animal development, from which apoptosis has emerged as an important protection against cancer.

Genetic Maps

At the beginning of the worm cell-mapping project specific genes could only be identified by a desperately laborious process. The method was to make mutants in model organisms like worms or fruit flies. By following inherited physical patterns, much as did Mendel in his peas, it's possible to track the proximity of mutated genes to each other. This gives a 'genetic map', more helpfully described as a 'linkage map'. Note, however, that such maps are just black boxes in a chromosome: they don't tell you anything about the genes themselves.

As he peered down a microscope at the intimate life of the worm it became clear to Sulston that the picture of how genes control development could not be complete without the corresponding sequence of DNA, the genetic material. The worm genome is made up of 100 million base-pairs, and in 1983 Sulston set out to sequence the whole thing, in collaboration with Robert

Waterston, then at the University of Washington in St. Louis. Genome sequencing had been launched by the aforementioned Fred Sanger with Alan Coulson, Barclay Barrell and their colleagues in 1977 when they produced the first complete sequence of a DNA genome. It came from a virus with a tiny amount of DNA – just over 5,000 bases. Then they sequenced human mitochondrial DNA (about 17,000 bases). This provided the foundation for complete sequencing of *C. elegans*, published in 1998, all 100 million base-pairs and about 20,000 genes, the first multicellular organism for which this feat was achieved.

Assembling the Toolkit

The step from genetic mapping to sequencing genomes depended on a number of technical advances. The first of these was made by Arthur Kornberg, working at Washington University in St. Louis in 1956, who isolated an enzyme from the bacterium *E. coli* that could copy DNA sequences – that is, using a DNA strand as a template, could link the building blocks of DNA (the A, C, G and T bases) to form nucleic acid polymers. You could say that this discovery launched the era of genetic engineering. As you might guess, numerous DNA polymerases from across the spectrum of organisms have since been identified. Each is involved in DNA replication or repair, and they all do essentially the same job: make a new DNA strand with a sequence complementary to the template.

The next items came from Werner Arber and colleagues in Switzerland and Martin Gellert at the National Institutes of Health in Bethesda, Maryland. They first detected enzymes in microorganisms that cut DNA at specific sequences that were named 'restriction sites'. These 'restriction enzymes' presumably evolved as an anti-virus defence but the great thing about them is that they can cut DNA from *any* organism and there are thousands of them that target different, short sequences. As a complement to restriction enzymes, in 1967 Gellert discovered an enzyme that joins DNA fragments together. So now we could cut and paste DNA.

A critical requirement for interrogating how a cell is behaving is to identify the messenger RNAs being made – because this gives a snapshot of which proteins are being made in a cell at any one time. It was therefore an immense discovery when, in 1970, Howard Temin at MIT and David Baltimore at the University of Wisconsin jointly isolated the enzyme reverse transcriptase (RT). This came from a study of retroviruses – viruses whose hereditary material is RNA rather than DNA. As part of their life cycle they turn their genomes into DNA that inserts into the host's genome – which gets reproduced (as RNA) to make more viruses. Hence, RT reverses part of the central dogma of molecular biology (DNA makes RNA makes protein) by using RNA as a template to make DNA. Viral RT works just as well on any messenger RNA and, in providing the means to convert mRNA back into DNA (a much more stable molecule), it has become an essential part of the molecular biology toolbox.

One piece was still missing from the armoury required for genome sequencing: a way of amplifying specific DNA segments, a gap that was filled in 1985 by Kary Mullis and colleagues when they invented the polymerase chain reaction (PCR). The capacity to make lots of copies of DNA fragments facilitates sequencing and cloning, and its availability meant that anyone could now play molecular biology, and they could do it with DNA from anywhere – themselves, yeast, tomatoes or whatever.

The only other thing you need is a vehicle – something to carry and propagate DNA segments. Step forward plasmids. Commonly found in bacteria, these are small double-stranded loops of DNA that are separate from the genome. They replicate independently and bacteria can be made to take them up from the outside. They can be split by restriction enzymes to provide a site into which you can paste a DNA sequence of choice using the joining enzyme. Once your modified plasmid (called a cloning vector) has been ingested, just leave the bugs to multiply then extract the plasmids from the bacterial culture to give you a permanent stock that can be frozen and fished out whenever you want to use the DNA sequence you inserted. It's called cloning, and each plasmid is a cloning vector. There's one other smart thing you can do with these clones. If you have included the right 'start' signals in the vector, the bacteria will use the plasmid as a template for its own transcriptional machinery. RNA will be made from plasmid DNA and that will be translated into the corresponding protein. A protein produced in this way is called a 'recombinant protein'. Insulin was the first recombinant protein to be made, revolutionizing the treatment of type I diabetes.

It's really worth bearing in mind that all these great discoveries on which the molecular revolution was built were just that: discoveries. They were revealed by excavating nature with much perspiration allied to a good deal of intuition. The tools required for the job were actually made for us by nature – and that explains the exquisite precision with which we can now manipulate DNA, genes and proteins.

The Sequencing of DNA

It was the wonderfully unassuming Fred Sanger who, having unveiled the sequence of amino acids in insulin, in 1980 worked out how to sequence DNA. For this he shared a Nobel Prize with Walter Gilbert, joining Marie Curie and the physicist John Bardeen as the only winners of two science Nobel Prizes. The basic idea of Sanger's method is to use a single strand of DNA as a template to make a complementary copy, just as happens when cells replicate their genomes. A DNA polymerase enzyme does the job provided it's supplied with nucleotides to be incorporated into the new strand. Fragments of about 500 bases are ideal for this method.

The brilliant bit, obvious of course after Sanger had thought of it, was to make slightly modified forms of each base so that another molecule cannot be added to the chain. In other words, the modified unit is a 'chain terminator'. Juggling the concentrations of normal bases and chain terminators leads to the production of fragments terminating at every base in the template. Now just incorporate fluorescent tags (four, giving a different light signal for each type of base) so that you can see the fragments, separate them by size and read off the sequence (Figure 4.3).

This method of generating labelled fragments of every size corresponding to a stretch of DNA and separating them by gel electrophoresis was used for the Human Genome Project. In 1992 John Sulston became head of a new sequencing facility, the Sanger Centre (now the Sanger Institute) in Hinxton, Cambridgeshire, home of the British component of the Human Genome Project – one of the largest international scientific operations ever undertaken. Despite the problems associated with integrating the work of many laboratories around the world, quite astonishingly in June 2000 President Bill Clinton was able to announce the completion of a 'rough draft' of the human

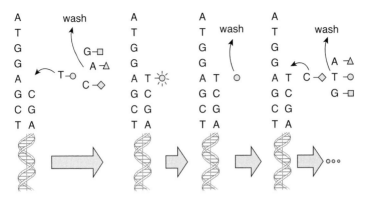

Figure 4.3   Sequencing DNA. The sequence to be determined is on the left (TCGAGGTA). This is a template and the matching sequence is made by DNA polymerase, adding one base at a time. Each of the four bases is tagged with a different fluorescent label. As each is added the light from the fluorescent label is measured, which gives the sequence.

genome. The sequence was largely completed by April 2003, two years earlier than planned, with approximately one-third (1,000 million bases) coming from the Sanger Centre. The complete human genome sequence, finished to a standard of 99.99 per cent accuracy, was published in *Nature* in October 2004. The work had been accomplished within both its budget and the original schedule.

In collaboration with Georgina Ferry, Sulston wrote *The Common Thread*, a compelling account of a project that has, arguably, had a greater impact than any other scientific endeavour.

As the Human Genome Project gained momentum it found itself in competition with a private venture aimed at securing the sequence of human DNA for commercial profit – that is, the research community would be charged for access to the data. Sulston was adamant that our genome belonged to us all, and with Francis Collins – then head of the US National Human Genome Research Institute – he played a key role in establishing the principle of open access to such data, preventing the patenting of genes and ensuring that the human genome was placed in the public domain. One clear statement of this intent was that, on entering the Sanger Centre, you were met by a continuously scrolling read-out of human DNA sequence as it emerged from the sequencers.

In the years since the first human genome was sequenced there have been utterly breathtaking advances in the speed and scale with which sequencing can be accomplished. Fully automated, high-throughput flow cells permit millions of DNA molecules to be read in parallel. Variously known as 'next-generation', 'second-generation' or 'massively parallel' sequencing, their power is quite astonishing. Nevertheless, the majority of sequencing methods still depend on measuring a light signal from a fluorescent tag. Yet more awe-inspiring technology is on the horizon in the form of 'third-generation' sequencing. One such method forces single strands of DNA in solution through small holes, the sequence being read from an electrical current flowing through the hole that is differentially affected by each of the four bases. Third-generation methods are still in their infancy but they offer the possibility of accurate sequencing without needing to amplify or label DNA. These astonishing technical advances have so accelerated the speed at which genomes can be sequenced and so lowered the cost that it is now feasible to obtain the complete sequence of someone's DNA within a few hours for a few hundred dollars, using a gadget about the size of a cell phone.

The power of sequencing is now delivering monumental projects like the Pan-Cancer Analysis of Whole Genomes, a collaboration of more than 1,300 scientists and clinicians from 37 countries, which has analysed 2,658 whole-cancer genomes and their matched, normal tissues across 38 tumour types. It's clear that over the next few years tens of millions of human genomes, both normal and tumour, will be sequenced that will reveal the underlying causes not only of cancers, but of the thousands of other genetic disorders. It's been evident for some time that ultimately these endeavours would converge on the target of characterizing the genomes of everything – that is, of all eukaryotic species. This has come to pass in the shape of the Earth BioGenome Project, which has been described as a moonshot for biology that aims to sequence, catalogue and characterize the genomes of all eukaryotes – organisms whose cells have a nucleus – on Earth over a period of 10 years.

With this vista in mind, we'll now look at the way in which cells regulate themselves and the steps that are most vulnerable to cancer-driving mutations.

5 What Is a Cell?

We have already encountered the seventeenth-century scientific genius Robert Hooke giving mankind his first glimpse of the miniature world of cells. It seems likely that *Micrographia*, his book of illustrations, came into the hands of one Antonie van Leeuwenhoek, the proprietor of a drapery business in Delft in the province of South Holland. In his spare time Leeuwenhoek had taught himself how to handle glass and he became so skilled that he was able to make the first compound microscope – one using two lenses that could magnify objects several hundred times. He started to look at animal cells and, in 1677, became the first person to see single red blood cells and sperm cells. He went on to discover bacteria, thereby making himself the first microbiologist. Through these technical advances, Leeuwenhoek changed the world by enabling scientists to look at individual cells, setting us on the road to understanding what we now simply accept as a fact of life – that all animals and plants are clumps of cells. Few outside science would recognize Leeuwenhoek's name today, which is slightly sad and somewhat ironic, given that his lifelong friend, the artist Vermeer, is world-famous.

The clumps that make human beings are, of course, quite large, variously estimated at about 50 trillion cells – that is, 50 followed by 12 zeros or 5×10^{13}. It's a bit vague, the range usually being given as 10^{13} to 10^{14} cells. You can define a cell as the smallest part of an animal that can survive on its own. That is, if you engage in a favourite pastime of biochemists by taking a bit of tissue, breaking it into single cells and putting them in a dish with some goodies, they often grow and even reproduce themselves. Look at them down a microscope and you will see sacs enclosed by a membrane – the plasma membrane. The sacs contain a gooey soup – the cytosol, in which float several smaller sealed sacs. The nucleus (containing DNA) and mitochondria (the 'powerhouses' of cells) are the most prominent. The whole lot inside the plasma membrane is sometimes called 'cytoplasm'.

The membranes that surround the internal sacs and the outer, plasma membrane are critical to cell behaviour because they control what passes in and out. They have evolved to be impermeable to more or less everything apart from gases and steroid hormones. This means that trans-membrane transfer, be it of information or nutrients, requires specific proteins that sit astride the barrier, acting as selective channels for molecules or ions – a biological border crossing. Membranes achieve this by being made of two layers of fat-like molecules. These are long and flexible, making membranes also extremely flexible, and that in turn enables cells to undergo prodigious contortions – they can squeeze through small gaps or down narrow tubes as, for example, blood cells do in the circulation.

These fatty molecules are in fact fatty acids – chains of carbon atoms with hydrogens attached (so they're 'hydrocarbons'). Pairs of fatty acids link to a kind of molecular bridge (glycerol) that has a phosphate group joined to a carbon atom, so the whole thing becomes a phospholipid (a lipid is something that's insoluble in water but readily dissolves in detergents or organic solvents such as chloroform or benzene). Two layers of these phospholipids sit tail-to-tail in membranes to make 'phospholipid bilayers'. Membranes also contain lipids with attached sugars and cholesterol. If you're wondering, the fatty acid 'tails' can be either saturated (no double bonds between carbon atoms) or unsaturated. And yes, before you ask, omega-3 fatty acids (with a double bond three atoms from their terminal methyl group) turn up in membrane lipids and lipid composition can affect things like trans-membrane signalling. But no, how much you happen to consume by way of omega-3 fatty acids isn't going to affect your chances of a heart attack.

The evolution from single-celled to multicellular organisms was accompanied by an increase in the repertoire of proteins that sit in or across the outer membrane – the pattern of expressed proteins determining the size, shape and location of each cell, together with the activities it carries out in fulfilling its job description. Imagine, then, a fatty bag sprinkled with a huge number of proteins that provide nourishment, news and sometimes direct contact with other cells – a boundary to the cell sometimes called a 'fluid mosaic membrane'.

Talking to Cells

A critical feature of multicellular organisms is that individual cells have to function as part of the whole – implying that they're sensitive to their environment. The decisions taken by cells – to divide or to change into a different type of cell or even to commit suicide – are enacted by the pattern and level of expression of genes switched on at any time. This is controlled by signals from the outside world transmitted across the plasma membrane and directed to the nucleus.

At the outset we defined cancer as 'cells behaving badly', and if that brings to mind an image of adolescents deaf to the world, it's not inappropriate because a feature of cancer cells is that they no longer respond to (chemical) signals in their environment. For a cell the most important message is whether it should divide to make two cells or stay as it is. It is corruption of that control by one means or another that lies at the heart of cancer. Environmental messengers are released from other cells in the body and travel via the blood circulation to their targets – signal receptor proteins on the surface of cells. The messengers are often protein hormones or neuro-transmitters (e.g., dopamine and serotonin). Perhaps the best-known hormone is insulin, made by cells in the pancreas and released when those cells sense the so-called well-fed state – raised blood levels of the sugar glucose. When insulin binds to receptors on liver cells the signal transmitted to the cell interior provokes a large increase in one type of membrane channel protein. These channels carry glucose into the cell where it tops up an energy store.

How does a messenger signal its presence without entering the cell? We've noted that proteins are flexible and that when something (e.g., a hormone) binds to its target site the 3D structure is altered. For a receptor this shape change is transmitted through the trans-membrane section to the internal domain (the blob at the other end if you like). There's a parallel in shaking hands with someone: your hand changes shape as you grip theirs. This form of signal transmission is often helped by hormones binding to two separate (but identical) receptors simultaneously, drawing them into an embrace. If the outer blobs are drawn together the cytosolic domains (inside the cell) have to follow suit – it's difficult to kiss while keeping your bottom halves far apart. Drawing the internal blobs together in turn causes them to change shape. The signal has been transmitted!

It's another facet of the incredible plasticity of proteins but a further question remains: how does the signal get from receptor to nucleus? The key lies in the fact that the internal receptor domains are, of course, proteins, but they are also enzymes. The shape changes driven by messenger binding switch on enzymatic activity. The enzymes are kinases – they transfer phosphate groups to a target molecule. The process is called phosphorylation, and the neat thing about receptor kinases is that they phosphorylate themselves on tyrosine amino acids. Hence, these receptors are known as receptor tyrosine kinases (RTKs). Activated RTKs are the first components of protein relays that carry messages from the cell membrane to the nucleus (Figure 5.1). Messages are relayed via proteins that are also enzymes – think of them as molecular Velcro: their sticky bits draw them together in an activation sequence that leads to phosphorylated proteins moving into the nucleus to activate specific programmes of gene expression. These cascades of enzymatic steps are switched off by other enzymes that remove phosphates.

For the idea of flexible molecules we should be grateful to the extraordinary polymath J. B. S. Haldane, who, in the 1930s when it was only just being shown that enzymes were in fact proteins, suggested that they stuck to their target by interactions that were weak in chemical terms. However, Haldane argued that, if contact was through several weak links, the enzyme might be able to distort the shape of its target, making it more reactive. His idea of flexibility in molecules, conceived long before X-ray crystallography was brought to bear, is central to how we think today not just about enzymes but about any interactions involving proteins. Haldane died of colorectal cancer at the age of 72. Once he knew what was wrong with him, he composed a poem ('I wish I had the voice of Homer ... ') that, as well as being funny, has advice about cancer relevant to this day.

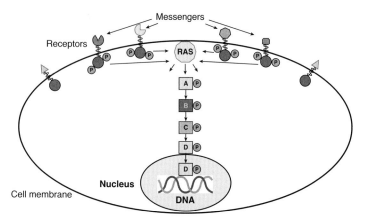

Figure 5.1 Cell signalling. Cells receive many signals from messengers that attach to receptor proteins spanning the outer membrane (RTKs). Activated receptors turn on relays of proteins. Receptors are shown as two blobs at either end of a wiggly line – the bit that crosses the membrane. RAS proteins are key nodes that transmit multiple signals. The shaded blocks represent a RAS-controlled pathway that 'talks' to the nucleus, switching on genes that drive proliferation. The arrows diverging from RAS indicate that it regulates many pathways controlling cell proliferation, cell differentiation, cell adhesion, cell death (apoptosis) and cell migration. Circled P = phosphate groups attached to the internal part of activated receptors; these act as launch pads for pathways that are activated by phosphorylation (e.g., A ➡ B ➡ C ➡, etc.). The dramatic effects of phosphorylation on protein shapes has nothing to do with size – phosphate groups on proteins are like flies on elephants – it's the (negative) electrical charge they carry that distorts the shape and hence modulates the activity of a protein. (A black and white version of this figure will appear in some formats. For the colour version, please refer to the plate section.)

We've spent a little time on RTK signalling pathways because these control the growth, location and lifespan of cells, and they are particularly prone to perturbation in cancers as a result of mutations. We should also mention that some signals act as negative regulators of cell growth. It's an example of the general principle that biological systems are usually a balancing act. Thus, at any time signals may tell cells to go forth and multiply, but others may instruct them to lie low, to remain quiescent rather than divide or even to commit suicide. The loss of negative growth signals – releasing a cellular brake – is also a significant factor in cancer progression.

A final complication in analysing cell signalling, let alone trying to modify it, is that the whole thing is a product of evolution – the random way in which cells try out different things to survive. It's like an old house in which the electrical wiring has been added to and fiddled with by many owners: new stuff is put in, but the old cables are left in the walls. Phone lines are disconnected but the wires remain. Humans are built in much the same haphazard way, so there's lot of old junk in our genomes and mostly it doesn't matter – but sometimes it's not fully disconnected. As Neil Shubin explained in his wonderful book *Your Inner Fish*, we're a bit prone to hiccups because the layout of our nerves that control breathing is a left-over from our fishy origins 400 million years ago – a layout utterly useless to us since our distant relatives clambered up the beach and got the hang of breathing without gills. But a misfiring neuron in our brain can turn on electric signals that control the regular motion of amphibian gills – a genetic recipe hoarded in the nuclear loft is inadvertently recalled. For the most part this result of evolution is no more than mildly embarrassing, although the poor fellow who made the *Guinness Book of Records* by hiccupping for 68 years may have used a stronger term. Mind you, hiccups is not always a useless irritation: about one in three people who have cancer of the oesophagus have frequent hiccups and difficulty swallowing. This may arise from a tumour growing into the trachea throwing the hiccup switch by mechanical pressure – thus setting off the involuntary spasm. But remember that we all get hiccups so don't panic next time it happens to you – unless it persists for a couple of days, then you might have an underlying problem.

Steroid Hormones

Before we leave the mechanics of signalling, a word about the other major group of signalling hormones – the ones that aren't proteins and are probably most famous for including 'performance-enhancing substances'. These are the natural steroid hormones, the most familiar being the sex hormones testosterone and oestrogen, vitamin D and cortisol. Made from cholesterol, which we encountered as a component of the plasma membrane, they are lipids – so they dissolve in fats. That enables them to pass through membranes and enter cells, where they can bind to specific steroid hormone receptors, the complexes then directing gene expression. Steroid hormone receptors are therefore transcription factors: they contain regions that bind to specific DNA sequences in the regulatory regions of their target genes. They are important in cancer because the large number of genes that respond, for example, to oestrogen includes several with critical roles in cancer.

Je Pense, Donc Je Suis un Blancmange

The key feature of all these pathways is that they connect the outside world to the nucleus of a cell. Evolution has generated thousands of proteins floating around in our cells that can be mapped into discrete signal pathways but, in the molecular jostle of the cell, each may affect any of the others – if not directly then via just a few intermediates. If you Google cell signalling pathways you get things that look far more baffling than maps of the Tokyo subway – zillions of dots joined by lines – so it's easiest to think of the cell as a blancmange: poke it anywhere and the whole thing wobbles.

Why is grasping this picture of what seems like a molecular madhouse important? We've acknowledged that, although the whole thing may look chaotic, cells somehow react to their environment by taking clear courses of action. But the reason for grappling with it at all, other than to be humbled by our ignorance, is that the signalling proteins are frequent targets for mutation, and these in turn have become a major focus for the development of anti-cancer drugs. Disruptions in these proliferation-controlling pathways, caused by mutations, are the targets for some of the contents of our drug cocktail cabinet, to be opened in Chapter 9, and it is helpful to have the image of cellular adaptability in mind when we confront the problem of the acquisition of resistance to anti-cancer agents.

Perturbing Cellular Balance

Compared to normal cells, most cancer cells have an increased tendency to pick up mutations as they multiply and make new cells – it's a characteristic called genome instability. One way of grasping this is to think about inherited cancers: they're rare (only about 10 per cent of all cancers) and they arise when an individual is born with a mutation in one copy of any of the genes involved in repairing damaged DNA. There's a high chance that after birth the second copy of that gene (allele) will be lost – and with it all function of that gene – resulting in genomic instability and inevitable tumour development.

In principle, disruption of any of the controls that normally regulate the cycle leading to cell division can produce genetic instability. DNA repair genes are an obvious target, but there are other mechanisms that monitor DNA integrity – notably various checkpoints that ensure all is well at each step as cells proceed around the cycle that leads to division. In addition, any of the components of signal pathways relaying information from outside the cell to the nucleus, and hence controlling cell division, are also potential targets for mutations that perturb the balancing act that is normal cellular behaviour. With all these points to aim for, it's not surprising that DNA sequencing has revealed that cancer frequently results from the combination of damage to multiple genes controlling cell signalling and cell division.

The Cycle That Makes Two Cells from One

Having noted that, starting from the fusion of two cells, human beings grow to 50 trillion, it's obvious that the single, fertilized cell has done a good deal of multiplication – variously called proliferation, replication or division. This process is commonly called the cell cycle (Figure 5.2) because one cell progresses through a series of steps that end up with it dividing into two identical daughter cells. The main steps in cell division comprise two growth phases (G1 and G2, G for 'gap', in which the cell doubles in size) that separate S phase (when the genome is replicated so that each daughter cell receives identical copies of DNA) from mitosis (M phase), when one cell becomes two. At key points – known as cell cycle checkpoints – conditions are assessed to determine whether it is appropriate for the cell to progress further through the division cycle.

After noting the importance of phosphorylation in controlling signalling pathways, it will come as no surprise that the central drivers of the cell cycle are kinases (enzymes that transfer phosphate to other proteins). Again, another cohort of enzymes, phosphatases, act to reverse the actions of kinases – that is, to knock off phosphate groups. So fundamental is this process to living systems that the major players have remained virtually unchanged during evolution. To find an ancestor common to yeast and humans you have to go back about

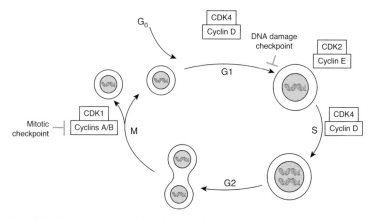

Figure 5.2 The cell cycle. DNA is duplicated in S phase, one cell becomes two in M phase and these events are separated by two growth (G) phases. DNA damage and mitotic checkpoints are marked. The main driving force for the cycle is the sequential action of a number of kinase enzymes that phosphorylate key targets. Kinase activity requires partner proteins called cyclins (cyclins D, E, A and B are shown) – so the enzymes are cyclin-dependent kinases (CDKs). Cyclins are made when required and then broken down: this ensures that the cell proceeds stepwise through the cycle and cannot go backwards or take shortcuts. Cells not actively dividing can adjourn to a resting, quiescent state (G_0) pending a signal to re-enter the cycle.

one billion years but you can swap yeast cell cycle genes for their human equivalents and vice versa and the cells still divide normally. So critical is the cell cycle – not only that it works when needed but that it is shut down when not – that there is an equally complex set of negative regulators – brakes. It's another example of a biological balancing act.

Major Kinase Targets in the Cell Cycle Clock

Of the checkpoints that monitor progression round the cell cycle, the most critical is that preceding the DNA replication step (S phase). At this point the retinoblastoma protein (RB1) acts as a master regulator by associating with a set of transcription factors called the E2F family. When RB1 binds to E2Fs, the complex *prevents* transcription of genes required for S phase.

Phosphorylation of RB1 releases its hold on the E2Fs: they become free to turn on their target genes – those needed for DNA replication and continued cell cycle progression (Figure 5.3).

A second major player in negative cell cycle control is the p53 protein – prosaically named from its mass. Its correct name is TP53 for tumour protein 53, which gives a clue to its importance in the cancer story. However, it's generally referred to as p53 and over many years it has emerged as central to the way cells respond to stress. p53 isn't necessary for normal cells to grow and divide – indeed it's so unnecessary that normal cells hardly bother to

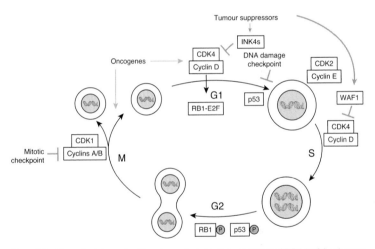

Figure 5.3 Overall picture of cell cycle control. Phosphorylation of RB1 in G1 releases transcriptional repression, enabling E2F proteins to activate key genes whose protein products drive cell cycle progression (*MYC*, *CDC2*, etc.). DNA damage activates p53: normal but not mutant p53 can then arrest the cell cycle and promote DNA repair after which p53 is switched off by phosphorylation. If DNA is not repaired p53 switches on genes that promote cell death. A further layer of control of CDKs and associated cyclins is by cyclin-dependent kinase inhibitors (CDIs) that include the *INK4* (inhibitor of CDK4) family and *WAF1*. RB1, p53 and the CDIs are known as tumour suppressors. Activating mutations in CDKs occur in some cancers, making them oncogenes. (A black and white version of this figure will appear in some formats. For the colour version, please refer to the plate section.)

make it. However, the cell machinery that detects DNA damage is exquisitely sensitive: it can detect just one break in the two metres of double-stranded DNA in a cell and respond by making p53. Thus, if you subject a cell to any kind of stress that damages DNA – even mild UV radiation in the form of sunlight – large amounts of p53 appear. The protein can protect cells against trying to propagate damaged DNA in two ways: (1) by arresting the cell cycle until the damaged DNA has been fixed; or (2) by killing the cell. It acts as a transcription factor, switching on genes, and a key target (ARF) halts the cell cycle by blocking CDK action. If DNA repair can occur during this arrest, p53 is switched off by phosphorylation and broken down. If not, p53 switches on a programme that targets mitochondria (the major source of energy) in the most effective way you could imagine – by making holes in the mitochondrial membrane. This torpedoing effect is fatal for the cell and p53 has switched on the best cancer protection you could have: the elimination of a cell with damaged DNA that could turn into a tumour precursor. This critical role led David Lane to describe p53 by the vivid phrase 'guardian of the genome'. Given its role, it's unsurprising that the most frequently altered gene in human cancers encodes p53. Over 50 per cent of human cancers carry loss-of-function mutations in this gene.

One further layer of cell cycle control is provided by a group of CDIs. These include INK4A and INK4B (INhibitors of CDK4) that block the kinase that phosphorylates RB1. By inhibiting this step, INK4A acts as a powerful cell cycle brake, acting in a parallel manner to ARF on CDK.

On first acquaintance the multiple layers of control that the cell cycle has evolved may appear incredibly elaborate, but they are to be expected when you consider the phenomenal precision with which organisms are built – cell division must be under the most rigorous control. It's clear that hyperactivation of any of the accelerators or loss of function of any of the cell cycle brakes might be prime movers in cancer, and indeed subversion of these controls by one means or another characterizes a cancer cell. When normal proteins become hyperactive the associated mutation(s) have, in the terminology, changed them from a 'proto-oncogene' to an 'oncogene', one of the two major classes of 'cancer genes'. The prefix 'onco-' means 'tumour', from the Greek *ónkos*, mass or bulk; 'proto-' comes from the Greek *prótos*, first.

In meeting RB1, p53, ARF and INK4 as regulators of the cell cycle, we have also encountered the leading members of the other main group of 'cancer genes' – known as 'tumour suppressors'. Despite the name, tumour suppressors have clearly not evolved to suppress tumours. They have normal roles in controlling cell division, acting as brakes when all is not well, so they might more accurately be thought of as 'growth suppressors'. When cells go off the rails we truly can, as our offspring are wont to do, blame it on their DNA. All of which brings us to the subject of mutations.

6 Mutations

In Chapter 5 we sketched the basic way in which cells process information from the outside world and the fundamental controls of the cell cycle. In principle, a mutation in any part that accentuates a positive or eliminates a negative could make a contribution to cancer. As a result of the amazing advances in DNA sequencing that we described in Chapter 4, we now know that most tumour cells accumulate tens of thousands of mutations, and so heterogeneous are tumours that every cell has a unique genome sequence. This implies that the longer you live, the more likely it is that a cancer will appear somewhere in your body – but the up-side is that two in every three only do so after the age of 60. Seventy-odd years ago the Finnish-born architect and amateur historian Carl Nordling drew a graph plotting the number of cancer deaths against the age at which they occurred. Result: a straight line. Remarkably, this holds true for pretty well every major cancer, regardless of country, sex or race. This linear relationship reflects the fact that cancers are mostly caused by a build up of specific events – mutations – on average at the rate of 1 every 10 years.

The accumulation of mutations drives aberrant cell proliferation – hence cancers take time to develop and are mainly diseases of old age. You might expect, therefore, that in animals the combination of size and how long they live would predict how likely they are to get cancers – but it doesn't. As first noted by epidemiologist Richard Peto, at the species level the incidence of cancer does not appear to correlate with the number of cells in an organism. For example, cancer arises in humans much more frequently than in whales despite a whale having many more cells than a human. On the other hand, humans are much less susceptible to cancer than are mice. Within species there is a positive correlation between the number of cells in an organism and cancer incidence, but across species this breaks down, as illustrated by elephants (average age 60, mass about 100 times that of a human), who rarely die of cancer – fewer than 5 per cent vs 11–25 per cent of humans – and the naked mole rat, which never, as far as we know, gets cancer. This is Peto's paradox and it carries the implication that animals have evolved innate anti-cancer defences of differing efficiency.

As we shall see, every upheaval you can imagine befalling a string of 3,000 million letters has been recorded and, as we survey the range of mutations, you might conclude that the astonishing thing is that cancers happen at all. In other words, the extent of genetic mayhem that frequently occurs leaves you wondering how a cell can actually work. Of course, a lot of DNA changes have no effect so they don't matter – and neither do those that are fatal for a cell. It will just die, the best form of cancer protection. That may sound like a high price to pay, but we lose about 50 billion cells a day in the normal course of events, so a few more to get rid of potential cancer cells is a good investment. Even so, the instability in their genomes means that cancer cells walk a slender tightrope between suicide and survival and their emergence is a rarity.

In the face of the information blizzard that emerges from tumour sequencing, the notion has evolved that we can indeed ignore most mutations because they are by-products that do not contribute to cancer development. These are often called *passenger* mutations, and one of the aims of sequence analysis is to sift these from the much smaller hand of *driver* mutations. Distinguishing *driver* mutations from the preponderance of neutral *passenger* mutations that characterize each cancer is not straightforward but relies on examining the type of genetic change and the affected genes.

As we noted at the end of the previous chapter, the two most obvious ways you could unleash proliferation from its normal controls would be to hyper-activate an accelerator or to lose a cell cycle brake. Either would fall into the category of *driver* mutations and it turns out that almost all cancers are driven by combinations of these two types of mutation. The current estimate is that of our 22,000 genes there are about 500 oncogenes and 100 tumour suppressor genes. This figure is slowly creeping up and a reasonable guess is that we're looking at about 1,000 'cancer genes' (Figure 6.1).

The First Experiment

We began our historical survey in Chapter 2 with the slightly less than serious notion that the story of cancer began in 1953 and, of course, it *is* about DNA and its encoded proteins. However, a Texas-born physician by the name of Peyton Rous, working at the Rockefeller Institute for Medical Research in New York at the beginning of the twentieth century, might reasonably claim to have done the first experiment in cancer molecular biology. He kept chickens in his backyard, and when one developed a large tumour he cut it out, mashed it up, filtered it through muslin and injected what passed through into normal chickens.

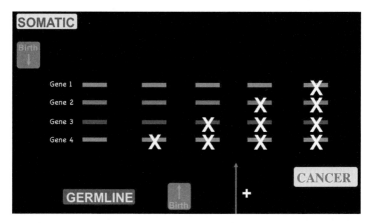

Figure 6.1 A mutational steeplechase leads to cancer. Of the tens of thousands of mutations that accumulate over time in a cancer cell, a handful of 'drivers' are critical (four are shown as danger [X] signs). Almost all mutations arise after birth (somatic), but about 1 in every 10 cancers start from a mutation present at birth (germline). Carriers are already one jump ahead and are much more likely to get cancer than those born with a normal set of genes. The rate at which mutations arise is increased by exposure to carcinogens (+) such as in tobacco smoke. (A black and white version of this figure will appear in some formats. For the colour version, please refer to the plate section.)

Doubtless to his surprise, these birds developed tumours looking exactly like the original. Rous, of course, hadn't a clue what he had done in molecular terms, but he was shrewd enough to comment in his 1911 paper that the agent carrying the tumour might be 'a minute parasitic organism'.

With that experiment Rous became a cancer immortal, sharing a 1966 Nobel Prize with Charles Huggins, who we met in Chapter 2, in the process creating a world record for the longest time between seminal discovery and prize-giving. The delay was another illustration of scientific advance being dependent on available technology – in this case the application of electron microscopy to biology whereupon the facility to magnify objects two million-fold revealed that Rous had injected his cancerous chooks with a virus. Viruses are about 20 times smaller than bacteria so it passed through muslin. And Rous was right: viruses *are* minute parasitic organisms.

It was then relatively easy to infect cells grown in the laboratory with Rous' virus and watch what happened. When normal cells are grown in culture dishes they fill up the available space and stop growing. However, cells infected with the Rous virus unstick from the dish and contract, literally pulling themselves together to begin dividing again. This makes their plasma membranes ruffle, giving a frilly appearance characteristic of a tumour cell. This process of turning normal cells into tumour cells is called 'transform-ation' and the question then became 'What was the virus carrying that gave it transforming power?'

The first thing to emerge was that Rous had isolated a retrovirus. We met these earlier with their capacity to reverse transcribe their RNA into DNA and insert it into the host's genome to reproduce it (as RNA) as lots more viruses are made. However, biology isn't perfect and, in hijacking the machinery of the cells sometimes odd bits of DNA are either lost or picked up by viruses as they go through this cycle. The Rous virus had picked up host DNA that encoded a complete protein – and it was this that caused the tumours in his chickens. These tumours were sarcomas, so the acquired gene was dubbed 'sarc'. This gene turned out to be present in pretty well all organisms and the human version is now denoted as *SRC* (pronounced 'sarc'). *SRC* encodes a tyrosine kinase, and the previous chapter described how important these are in cellu-lar control. The *SRC* gene acquired by the virus had suffered slight damage – it had lost a regulatory region that normally shuts down the protein's kinase activity. Hence the viral kinase is always switched on, so it continuously drives cell proliferation. Thus was the first oncogene revealed and it's something of a surprise that in humans oncogenic *SRC* has turned out to be rare, although some colon cancers have almost exactly the same mutation as the Rous sarcoma virus.

The Age of Oncogenes

The stage was then set for Michael Bishop to dissect another virus, MC29, that causes carcinomas in birds. As these are by far the most common type of human cancers, Bishop reasoned that MC29 might be the carrier of something interesting. It was a brilliant example of educated guesswork because MC29 had acquired a gene that was named *MYC* ('mick') – *MYC* because the virus causes tumours in *my*eloid *c*ells, a type of bone marrow cell. Subsequently, *MYC* has been revealed as the most frequently activated oncogene in human cancers – most if not all tumours have abnormal MYC expression. It's difficult to resist quoting Bishop's words: 'By means of accidental molecular piracy … viruses may have brought to view the genetic keyboard on which many different causes of cancer can play, a final common pathway to the neoplastic phenotype' – neoplastic meaning tumour-like and phenotype being an observable characteristic.

Gradually, more and more viruses were found that had picked up genes from their hosts, converting them into cancer-causing genes in the process. They turned up in every organism, from humans to flies to worms, and this high level of conservation across species was a clear indication that their encoded proteins were involved in controlling cell proliferation. This was a critical moment in cancer biology; for their discovery of the cellular origin of retroviral oncogenes, Michael Bishop and Harold Varmus shared the 1989 Nobel Prize.

The First Human Oncogene

The evidence that viruses could cause cancer due to genes picked up from their hosts was of course a 'smoking gun', but the crucial piece of evidence took a while coming – until 1983 in fact. The experiment was simple in design: extract DNA from a human tumour, chop it into short lengths and add them to cells growing in culture. The pieces of DNA are taken up by the cells, some of which start to divide abnormally and form clumps, as we described just now. Take cells from the clumps, inject them into mice, and when a tumour grows remove it and isolate the gene responsible. By showing that the cancer-promoting gene came from the original human tumour DNA, Robert Weinberg, Channing Der and their colleagues identified the first human oncogene, *RAS*. It transpired that the gene had already been identified – in a retrovirus – and named because the virus causes *rat* sarcomas.

We now know that the human genome contains three closely related *RAS* genes (*KRAS*, *HRAS* and *NRAS*). As RAS has a central role in cell signalling, it's no surprise at all to find that mutations generating oncogene from proto-oncogene occur in about 20 per cent of human tumours.

Making Mutant Proteins

As we've observed, the genetic chaos of cancer cells can feature every imaginable kind of DNA disruption, but they can be grouped into five broad types: (1) single base change; (2) losing chunks of a gene; (3) shifting gene segments to abnormal locations; (4) duplicating whole chromosomes or even the entire complement of chromosomes; and (5) losing whole chromosomes. To get the hang of what mutations do in cancer, it's helpful to look at specific examples of each of these groups to see the upshot and to note how particular types of mutation associate with different forms of the disease. So, we're going to open the book of life – the string of 3,000 million base-pairs that is your DNA which, written in the same font as this book, would run to 120,000 pages of what looks like the most boring tome imaginable. However, it's the detail that turns this volume into a sensation. Before we start, recall that the DNA code comes in groups of three bases. As an example, in RNA made from this code (by transcription) the triplet GGG inserts the amino acid glycine into the growing protein: mutate a G to a T (to give GTG) and a different amino acid (valine) goes into the protein. For triplets to be read they must follow the first codon of a messenger RNA (the start codon) in blocks of three. The start codon always codes for the amino acid methionine, which is why sequences start with ATG.

A Single Base Change: Minimal Mutations in Molecular Switches – RAS

1.   ATGACTGAATATAAACTTGTGGTAGTTGGAGCTGGTGGCGTAGGCAAGAGTGCCTTGACGATACAGCTAA

2.   ATGACTGAATATAAACTTGTGGTAGTTGGAGCTGTTGGCGTAGGCAAGAGTGCCTTGACGATACAGCTAA

At a quick glance can you spot any difference between these rows? No? Excellent! Your first go at sequence analysis and you got it wrong! No matter, unless you have one of the 20 per cent of cancers driven by mutation in *RAS*, in which case you might wish to look more closely at the first 70 bases of the *RAS* gene. Count triplets until you get to the 12th: the top line is GGT but the bottom is GTT – that's mutant *RAS* (Figures 6.2 and 6.3). How can a one-letter change be a potential death sentence? The mutation replaces the smallest amino acid glycine (a single hydrogen atom as its side-chain) with valine (a branched side-chain, non-polar amino acid).

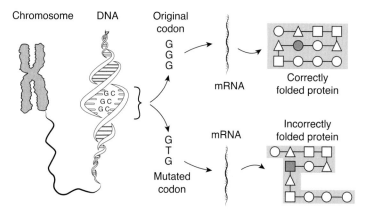

Figure 6.2   Making a mutation. Changes in the sequence of DNA may change the amino acids in a protein. The sketch shows the effect of the smallest change – a single base in one codon (called a 'point mutation'). The normal codon (G**G**G) encodes the smallest amino acid (glycine), the mutated codon (G**T**G) encodes a much larger one with different chemical properties, the effect of which is to change the shape and action of the protein.

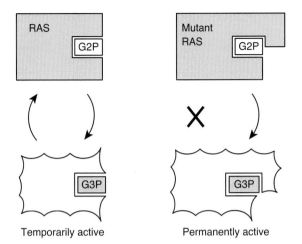

Figure 6.3 Normal and mutant RAS. A single amino acid change locks this molecular switch in the 'on' position. The conformational change means the RAS enzyme cannot convert GTP to GDP, and the signal pathway it controls is permanently active. The conformational change means the RAS enzyme cannot convert G3P (GTP) to G2P (GDP), and the signal pathway it controls is permanently active.

As we've noted, RAS plays a critical role in cellular relays talking to the nucleus because it works like a switch: it's either 'off' or 'on'. The switch is 'on' when RAS binds to a small molecule called GTP. GTP is one of the four building blocks of DNA but here nature recruits it for a completely different job – changing the shape of RAS to make available sites that activate downstream signalling molecules. It's another example of evolution by trial and error – using whatever is at hand. Unintelligent design one might say.

RAS turns itself off because, as well as being able to talk to other proteins, RAS is an enzyme. It chops a phosphate off GTP (guanosine triphosphate) to yield RAS bound to guanosine diphosphate (GDP). The effect of the glycine–valine swap is to change the delicate 3D structure of the protein, preventing this dephosphorylation. RAS remains locked in an embrace with GTP, the molecular switch is permanently 'on', and so too are the signalling pathways it instructs. Precisely the same amino acid change (in haemoglobin rather than RAS) underlies sickle-cell disease.

Missing Bits: Deaf to the World – EGFR

1. ATGCGACCCTCCGGGACGGCCGGGGCAGCGCTCCTGGCGCTGC-
    TGGCTGCGCTCTGCCCGGCGAGTCGGGCTCTGGAGG ...
2. GAGG ...

Our next coding quiz is a bit easier. There's a big segment missing from the second sequence – if that encodes a receptor perhaps it's lost the part that's normally outside the cell providing a sticky binding site for a messenger. Congratulations sequence analyst! This sequence information *is* for a receptor, one to which a messenger protein called epidermal growth factor (EGF) binds, hence it's the epidermal growth factor receptor (EGFR). Losing the external domain has an unexpected effect: it allows the internal parts of receptors to come together, change shape and turn on downstream signalling (Figures 6.4). Receptors are permanently activated without needing external prompting. It's a great example of a mutation side-stepping normal cellular controls, and it's an important driver in some forms of breast and lung cancer.

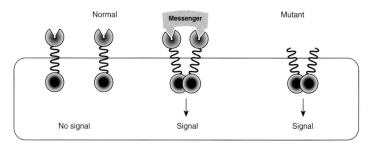

Figure 6.4 Signalling by normal and mutant receptors. Loss of the external domain can activate the intracellular signal without a messenger.

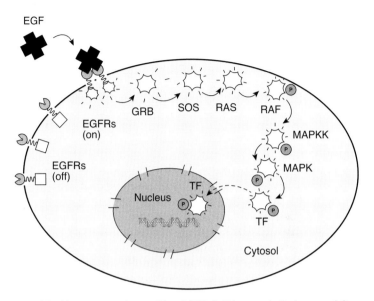

Figure 6.5 Message to nucleus: getting at DNA. In this example the hormone (often called a growth factor) is epidermal growth factor (EGF), which binds to its receptor (EGFR) and switches on a relay of proteins, called the RAS–MAPK pathway, to carry its signal from the membrane to the nucleus (MAPK = *m*itogen-*a*ctivated *p*rotein *k*inase: 'mitogen' is another name for growth factor). Two small proteins, GRB and SOS, link the receptor to RAS. Each circled P represents a phosphate group attached to a protein. TF = transcription factor.

EGF is one of many growth factors that turn on cell signalling pathways of the sort we generalized in Chapter 5. Figure 6.5 shows one of these – the MAP kinase pathway. It is turned on when the EGF receptor is activated: this in turn throws the RAS switch and activates RAF – the next protein in the relay that we'll come back to shortly in the context of skin cancer. The rest of the pathway leads to the movement of transcription factors into the nucleus. RAF is followed by MAPKK and MAPK, which form an enzyme cascade. The MAPK pathway is conserved across all organisms as a central mechanism for signal transduction.

Each amplifies the signal, leading ultimately to a change in the pattern of gene expression – and hence the behaviour of the cell.

Patching Proteins: Chromosome Translocations Make Novel Proteins

1.   … GCTGTCCTCGTCCTCCAGCTGTTATCTGGAAGAAGCCCTTCAGCGGCCAGTAGCATC …

2.   … CGATGGCGAGGGCGCCTTCCATGGAGACGCAGATGGCTCGTTCGGAACACCACCTGG …

3.   … CGATGGCGAGGGCGCCTTCCATGGAGACGCAGAAGCCCTTCAGCGGCCAGTAGCATC …

Trickier still! Now you have two different sequences (1 and 2). The third is a mutation: the front of 2 has shifted to join the back of 1 – it's called a chromosome translocation and these are particularly common in cancers arising in blood and immune cells, with different translocations being associated with particular types of leukaemia. It's remarkable that chromosome translocations have any effect because the chunks of DNA that are brought together from distant parts of the genome have to be joined so that the new sequence can be read as triplets if a new protein is to be made. Work they do, however, because if you take one of these human 'fusion genes' and express it in mice or zebra fish they get the same type of leukaemia, confirming that the completely novel protein – a result of DNA rearrangement – is sufficient to trigger the cancer.

The most familiar chromosomal translocation occurs in over 90 per cent of chronic myeloid leukaemia (CML) cases and in 10–20 per cent of acute lymphoblastic leukaemias (ALLs). Often called the Philadelphia chromosome – because that's where Peter Nowell and David Hungerford discovered it in 1960 – it comes about when a piece of chromosome 9 displaces part of chromosome 22. It was Janet Rowley who showed that this DNA shuffle moves a bit of a gene called *BCR* and puts it directly in front of a gene called *ABL1*. ABL1 is a close relative of SRC and it too is a tyrosine kinase enzyme. The effect of having BCR stuck on its front end is to switch it on permanently – it's just a different kind of activating mutation. This one makes the BCR–ABL1 fusion protein – the driving force for leukaemia development.

Revelations from Leukaemia

The leukaemias have been the source of much information about cancer development through the use of fluorescent probes that can identify specific genes in single cells (a method called FISH – f**luorescence *in situ* h**ybridization). This means you can show that in normal cells *BCR* and *ABL1* are on chromosomes 22 and 9, but in leukaemic cells they are on the same chromosome. What's more, it's now possible to sequence DNA from an individual cell to give the complete mutational pattern for that cancer. Applying FISH to childhood ALL has shown that in three-quarters of cases there is a translocation that fuses the two genes *TEL* and *AML1*. This event occurs in one fetal cell that then multiplies. Although ALL is the most common childhood malignancy (30 per cent of all cancers in children under 15 years of age), it's still very rare, about 47 cases per one million per year in Europe. Remarkably, whereas about 1 child in every 100 has a *TEL–AML1* fusion, only 1 in 100 of them gets leukaemia. This tells us that *TEL–AML1* fusion, in contrast to *BCR–ABL1*, is a 'promoter' event, following which further mutations are required – a vivid case of genetic roulette. Consistent with this requirement, ALL appears at widely varying ages, from birth up to about 14 years.

The approach of tracking DNA in single leukaemic cells has revealed a striking and central feature of cancer: different cells have distinct patterns of mutation, a property we alluded to in Chapter 1. This tells you that tumours are made up of distinct groups of cells, that is clones, even though they all started from the same fusion event in one cell – *TEL–AML1* in the case of ALL.

Following individual genes in this way yields 'genetic snapshots' of astonishing diversity. Thus, a clone may duplicate a gene and subsequently lose that extra copy as it evolves into a different, closely related, clone. Mel Greaves, a leading authority on childhood leukaemia, has coined the term 'clonal architecture' to describe the heterogeneity that arises when a single cell with its hand of mutations evolves into a mutational mixture of individual clones present in different proportions, each with its own capacity for self-renewal.

One implication of tumour heterogeneity is that treating cancers with drugs will be a nightmare because you're trying to hit lots of targets rather than one, and the targets are changing all the time – a problem we'll come to in Chapter 8. Liquid tumours are particularly amenable to this type of study, but it's now clear that solid tumours also comprise multiple clones, independently modulating their mutational profile and proliferating at different rates. As we remarked in Chapter 1, cancers are a form of dynamic Darwinism, and in trying to deal with them we're taking on the might of evolution.

Replacing the Controller

In addition to generating novel proteins by fusion, chromosome rearrangements can produce cancer drivers by placing a normal gene under abnormal control – that is, DNA shuffling places a different promoter in front of a gene so it is under new control. The upshot is that a normal protein is made but it may be in the wrong type of cell or in the right cell at the wrong time. An example occurs in Burkitt's lymphoma (of which more later): the gene involved is *MYC*, which finds itself regulated by a DNA sequence that usually controls a completely different gene without disruption of its protein-coding region. The new promoter normally controls immunoglobulin genes – continuously expressed in B cells – and the sustained transcription of *MYC* ultimately leads to cell transformation.

Multiplying Genes

    MYC:

        CTGGATTTTTTTCGGGTAGTGGAAAACCAGCAGCCTCCCGCG-
        ACGATGCCCCTCAACGTTAGCTTCACCA ...
        CTGGATTTTTTTCGGGTAGTGGAAAACCAGCAGCCTCCCGCG-
        ACGATGCCCCTCAACGTTAGCTTCACCA ...
        CTGGATTTTTTTCGGGTAGTGGAAAACCAGCAGCCTCCCGCG-
        ACGATGCCCCTCAACGTTAGCTTCACCA ...

Chromosome translocation is one means of hyper-activating MYC but, given that we now know it is essential for cell proliferation, you might guess that in cancer alternatives have evolved. Indeed, and it's now clear that most types of cancer overexpress MYC (i.e., have multiple, identical copies of the *MYC* gene as shown above). In some colon cancer cells, for example, there are over 100,000 molecules of MYC, whereas normal cells make fewer than 1,000. Perhaps the most drastic mechanism is the duplication of entire genes. Occurring frequently in human cancers, it was first discovered in the most common solid tumour in children – neuroblastoma. In this disease the *MYC* family member *MYCN* can be amplified to the level of hundreds of copies per cell. This happens in ~25 per cent of cases and correlates with poor prognosis.

Other examples of gene amplification reflecting probable cancer progression include 20 per cent of primary breast cancers in which one of two receptors on the cell surface are amplified, EGFR or its close relative HER2. This results in excessive receptor numbers, and HER2 amplification is an especially bad sign because these tumours are usually particularly 'aggressive'. As you might guess from the role of MYC in proliferation, an even worse scenario is high levels of both HER2 and MYC, which happens in about 12 per cent of breast tumours. HER2 is the target of Herceptin, an effective treatment for about 20 per cent of early-stage breast cancers that we'll return to later.

Given the importance of tumour dissemination (the overall five-year survival rate in the USA is 90 per cent, but is below 30 per cent for cancers that have metastasized), an obvious question is whether there's an increase in mutation load with progression from early to metastatic tumours. A large-scale study published in 2019 showed that the mutational load does increase in meta-static tumours – on average from 2.4 to 3.8 mutations per million bases.

Like all molecular factors in cancer, there's nothing odd about gene amplification – it's a normal part of evolution. Quite large stretches of human DNA are frequently duplicated to the extent that about 5 per cent of our genome comprises 'segmental duplications' – long bits of DNA (typically over 1,000 bases) with nearly identical sequences (90–100%). This repetitive stuff makes up over half the Y (male) chromosome, and comparison of the genomes of any two people will reveal an average of about 80 genes that are present in different numbers. That's called 'gene dosage': it's reflected in the number of proteins each gene makes and is a significant contributor to the differences between us.

In the last few years it's emerged that, distinct from gene amplification, duplication of the whole genome (i.e., all chromosomes) is relatively common in cancers. Thus, for example, whole-genome doubling (WGD) was detected in almost 30 per cent of nearly 10,000 advanced cancer patients, its prevalence varying by cancer sub-type. Non-small cell lung cancer (NSCLC) has one of the highest frequencies of WGD, and a project called TRACERx (Tracking Cancer Evolution through Therapy), set up in 2014, aims to follow 840 people being treated for NSCLC from the time of diagnosis to reconstruct the tumour's evolutionary history. Nearly three-quarters of the first 100 tumours analysed had WGD. Chromosome deletions and duplications and point mutations in DNA were also prevalent. These occurred as a result of tobacco exposure, and WGD appears to be an early event in lung cancer development in smokers. It seems probable that WGD provides a buffer of genes essential for tumour survival against a high background of genetic disruption, and there is evidence that cancer cells with a duplicated genome can indeed survive in the presence of many more mutations than they could otherwise tolerate. Predictably, WGD is associated with poor prognosis across cancer types.

Genes Go Missing: *RB1*

RB1:
ATGCCGCCCAAAACCCCCCGAAAAACGGCCGCCACCGCCGC-
CGCTGCCGCCGCGGAACCCCCGGCACCGCCGCCGCCGCCCC-
CTCCT ...
RB1: . . . .
. . . .
. . . .
. . . .
. . . .

This example of a mutation is clearly the flip-side of amplification – a gene that we encountered as a 'growth suppressor' in the last chapter, *RB1*, is lost. In cancer the complete loss of large segments of DNA, or indeed entire chromosomes, is far from a rare event. If this removes tumour suppressors such as *RB1* or *TP53* it would provide a powerful cancer-promoting event.

Robert DeMars had been a graduate student of the legendary Salvador Luria, a Nobel Prize winner for his work on viruses that infect bacteria. He worked for many years at the University of Wisconsin, Madison and it was he who first made the suggestion, in 1969, that cancers could be caused by losing something. In 1971 Alfred Knudson gave substance to this idea by invoking retinoblastoma, a rare inherited childhood disease in which tumours develop in the eye. Retinoblastoma comes in two forms: sporadic, in which there's no family history and just one tumour develops in one eye: and inherited, in which tumours arise in both eyes. Knudson, noting that we normally have two copies of each gene (alleles), proposed that you only got retinoblastoma when both copies of a gene are knocked out. There are two forms because children inherit either one defective allele and then lose the other (familial) or are born genetically normal but pick up mutations in both genes. Either way, to get the disease you need to have both copies knocked out (Figure 6.6).

This became known as the 'two-hit model' as it needed two genetic events – and remarkably perceptive it was, given that it was based on no molecular evidence whatsoever. It's transpired to be absolutely correct for retinoblastoma, although it was a very long journey from Knudson's idea to the identification and cloning of the *RB1* gene. We now know that the *RB1* gene is completely defective in all retinoblastomas but, given its role in the cell cycle, it's no surprise that *RB1* is knocked out in a good many other tumours, including 20–30 per cent of lung, breast and pancreatic tumours. *RB1* is therefore the classical model for a 'tumour suppressor' gene. *Both* copies of the gene must be inactivated for the tumour to develop (geneticists call this

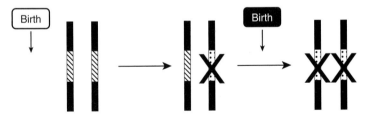

Figure 6.6 Retinoblastoma. This cancer occurs only when both copies of a gene (*RB1*) are knocked out (X). In one form of the cancer the individual is born with two normal copies of the gene and both become mutated after birth (they're somatic mutations). In the other (inherited) form, children are born with one defective gene (black 'birth' box) and lose the other copy by somatic mutation. (Mutations acquired by any of the cells of the body except the germ cells, sperm and egg, are somatic mutations: that is, they are not passed on to children.)

behaviour 'recessive'). The loss of just one copy has no effect. This contrasts with oncogenes in the cell cycle, which are 'dominant' because mutation of just one allele *is* sufficient for an effect to be seen.

Tumour Protein 53

TP53, together with the retinoblastoma protein, are major negative regulators of cell cycle progression. The *TP53* gene is often completely lost in human cancers but, in addition, over 6,000 mutations have been found in p53. Across the spectrum of human cancers over 70 per cent have mutations that affect p53's activity. Usually these are not inherited (i.e., they're somatic) and they are particularly common in all types of lung cancer, in breast tumours and in brain tumours. As you might expect, mutations that bring about changes in p53 are often associated with the activation of oncogenes – loss of brake and gain of accelerator.

p53 first came to light 40 years ago in studies of a virus that causes tumours in some animals. When the responsible protein was isolated it came with p53, a cellular protein, stuck to it. Eventually it became clear that the viral protein was destroying p53, enabling the virus to reproduce at will. However, it was only with experiments in the 1990s by Allan Bradley's group in Houston using transgenic mice that the role of p53 became clearer. When both alleles were knocked out so that the mice were unable to make any p53 at all, to the surprise of almost everyone the mice developed perfectly normally to adulthood and could even breed. Fortunately, Bradley's group persisted by keeping the transgenic mice, while wondering why a protein, so important that it was destroyed by a tumour virus, could be dispensed with at seemingly no cost. Over the next six months their patience was rewarded as it gradually emerged that these otherwise normal mice were extremely prone to cancer. So much so that, after about six months, three-quarters of the mice with no p53 had developed tumours of one sort or another. What had happened was perfectly consistent with the role of p53 in the cell cycle: it isn't needed for normal cell division, but has evolved as part of a major mechanism for protecting cells from replicating damaged DNA – which accumulated over time in the transgenic mice.

Thus, *TP53* is generally considered to be a tumour suppressor gene but in fact its behaviour is less clear-cut than that of *RB1*. The inheritance of one defective *TP53* allele is a rare event, but it gives rise to a disease called Li–Fraumeni syndrome in which 50 per cent of the mutant allele carriers develop diverse cancers by 30 years of age, compared with 1 per cent in the normal population. In this case, in contrast to the recessive behaviour of *RB1*, *p53* is acting in a dominant negative manner (i.e., mutation in one allele yields a protein that affects the activity of the protein made from the normal allele). In passing we should note that it is relatively rare for people to develop more than one simultaneous cancer: those with an inherited disposition are the exception.

The Double Life of p53

As noted in the previous chapter on the cell cycle, p53 acts as a transcription factor. We've emphasized the importance of shape for protein activity, and X-ray crystallography has given us a clear picture of how 'fingers' of amino acids stick out from p53 and slot into regulatory sequences in the grooves of the DNA double helix (Figure 4.1), providing a target for the start of gene transcription. Over 6,000 mutations have been found in p53: unsurprisingly, the 'fingers' that interact directly with DNA are mutational 'hot spots'. The first direct evidence that cigarette smoke causes cancer came from mutations in this region caused by benzo[a]pyrene, one of the many carcinogens in tobacco smoke – a specific effect on a key tumour suppressor.

Exploding DNA

We began this chapter by commenting that DNA has been the hapless recipient of just about every type of damage you could imagine. Despite the variety of mutations, however, the conventional picture of cancer is of tumours, or at least clumps of tumour cells comprising a clone, evolving through the progressive accumulation of mutations over extended periods of time, albeit in a random fashion. However, having marvelled at the range of mishaps that can befall our genetic inheritance, the thought might have occurred that, rather than simply plodding along the malignant road, once in a while something catastrophic might happen – a kind of genetic eruption giving cancer evolution a massive boost.

I think it's true to say that such a thing didn't occur to anyone until, in 2011, Philip Stephens and colleagues at the Wellcome Trust Sanger Institute near Cambridge showed that just such a cataclysmic event could happen. It takes the form of a number of chromosomes shattering into fragments – probably as a result of something drastic going wrong when a cell is in the act of dividing. The DNA repair systems react by going into overdrive, frantically gluing the bits of DNA together as best they can. This seems to be a fairly random business and in most cases the result will probably be cell death – its genome is so damaged it can't go on. Infrequently, however, the cell does survive, now a giant step further on the way to cancer.

The process has been called *chromothripsis* (the Greek *thripsis* meaning shattering into pieces) and, although it happens in only a few per cent of cancers, it's not confined to particular types and has been found in leukaemia, skin and lung cancers and a variety of other tumours.

The DNA repair systems that spring into action in chromothripsis are also critical in routine cellular housekeeping. DNA in a normal human cell suffers an estimated 20,000 DNA hits – that is, damaging chemical reactions – every day. A variety of DNA repair processes correct nearly all these hits to the extent that an average of less than one is 'locked in' – that is, passed on when a genome is replicated. One upshot of this is that every time human DNA is passed from one generation to the next it has accumulated new mutations. Several hundred genes are devoted to DNA repair and these are an important sub-class of tumour suppressor. Mutations impairing their activity can lead to genetic instability. These DNA damage events can occur at high frequency; for example, a single mammalian cell accumulates ~10,000 abasic (apurinic) lesions.

The life-shortening genetic disorder xeroderma pigmentosum (XP) arises from a mutation that limits the body's capacity to repair DNA damaged by ultraviolet light. Sufferers are prone to skin cancers and have to avoid sunlight completely as there is no treatment. In addition to XP, several other conditions predispose individuals to cancer as a result of DNA repair defects, including Werner syndrome, Bloom syndrome and Fanconi anaemia. The rare, inherited neurodegenerative disease ataxia telangiectasia (AT) that carries an increased risk of cancers, particularly leukaemia and lymphoma, is also often included in this category. AT is caused by mutations in a protein that detects double-strand breaks in DNA: its inactivation prevents activation of the tumour suppressor p53.

Micro RNAs

So far we've seen the effects on the two major classes of 'cancer genes' – oncogenes and tumour suppressors. However, we need a word about a third group that adds a level of weirdness to the story because its members can behave as either oncogenes or tumour suppressors. What's more, they give a different twist to the central dogma 'DNA makes RNA makes protein' by stalling halfway – that is, their genes are transcribed into RNA but the RNA isn't messenger RNA and it doesn't make a protein.

Meet micro RNAs, genes that make short RNAs – just 18–24 bases in length. Their role is to block the production of proteins: their short sequences are complementary to sequences in bona fide messenger RNA, enabling them to form pairs with mRNA bases, much as DNA forms base-pairs in the double helix. That's unusual because, in contrast to DNA, RNA doesn't form double strands. The best RNAs can manage is for short regions to curl up on themselves and form loops. However, when micro RNAs bind to their target mRNA, the effect is to inhibit synthesis of a specific protein. The block may be temporary or it may lead to mRNA breakdown. Either way, the upshot is that synthesis of a specific protein is inhibited.

There are over 800 micro RNAs in the human genome, and they play roles in all the things that matter for a cell: growth, death and the process of developing from one cell type to another. It follows that micro RNAs are important in cancer: the pattern of expression of many changes in tumours by comparison with the corresponding normal tissue. Quite often the level of groups of micro RNAs is lower in tumours, suggesting that switching on those genes gives protection against tumour develop-ment – they act as tumour suppressors. Thus, for example, p53 can regulate transcription of micro RNAs to drive cell cycle arrest and apop-tosis. On the other hand, micro RNAs that target tumour suppressor mRNAs can be said to be oncogenes. Their expression patterns vary markedly between different tumour types, but it's possible, for example, to distinguish leukaemias from solid tumours by their micro RNA profiles.

Genetic Variations

It's widely known that human beings 'share' the same DNA. Well, almost – if we're 99 per cent identical to chimps and share half our genes with bananas, there's not much scope for variation among ourselves. But there is just enough – 0.1%, one base in every 1,000 – to make us different from one another. By and large this variation doesn't affect our capacity to function, with the exception of inherited mutations in specific genes that drastically affect their activity (e.g., that causing cystic fibrosis).

An individual's genetic variants are identified by comparing their DNA sequence with that of a reference genome; so far over 300 million have been found. The biggest group of these minor DNA variants are single-nucleotide polymorphisms (SNPs, pronounced 'snips'). SNPs are exactly what their name says: a difference of a single nucleotide (base) in two otherwise identical stretches of DNA sequence. A given SNP may be present in one allele (on one chromosome) but not in the other within one individual, or it may be in both alleles or neither. As a rough guide, a SNP is a DNA variant detectable in over 1 per cent of the population. Most SNPs do not affect protein sequences: those that do change a protein amino acid sequence are called non-synonymous SNPs.

SNPs have gained prominence in the context of breast cancer, of which 15–30 per cent are driven by inherited genetic alterations. In addition to *BRCA1* and *BRCA2*, a number of other genes (*PALB2*, *ATM* and *CHEK2*) are known to acquire mutations in breast cancers, but these account for only 30–40 per cent of familial cancers. The hunt for the rest of the heritability link has prompted genome-wide association studies (GWAS) that have uncovered several hundred SNPs, currently accounting for about 18 per cent of breast cancer heritability. The point to grasp is that each of these variants on its own has a negligible effect, but appropriate combinations can exert a significant 'driver' effect. A one-line summary would be that the genetics of breast cancer is highly complex, and we'll see, literally, just how complex when we come to treatment.

Viruses

We were unable to avoid mentioning viruses in talking about warts and catching cancer. Perhaps it's not surprising that they've already crept in as there's an estimated 10^{31} virus particles in our biosphere. We'll come shortly to bacteria that outnumber the cells in the human body by about three-fold but, all told, viruses outnumber them by 10 to 1.

Viruses come in a bewildering variety of types and sizes, but they all have a core of genetic material with a protein coat (called the capsid). Some also have a membrane envelope much like the cell plasma membrane. Their most important feature, however, is that they are cellular parasites. They survive by sticking to cells and then elbowing their way in to hijack their host's molecular machinery to reproduce themselves. There are two main categories: those with DNA as genetic material and those with RNA.

DNA Viruses

DNA viruses are responsible for quite a few human miseries, including the common cold, herpes, smallpox, polio and chicken pox, as well as its derivative, shingles. Given the amazing diversity of viruses and the fact that their lifestyle involves throwing something of a molecular spanner into the works, it is perhaps surprising that they are not a major cause of human cancer. Collectively, infectious agents that include viruses and bacteria cause about 15 per cent of human cancers.

The capacity of human papillomaviruses (HPVs) to cause cervical cancers arises from viral proteins knocking out p53, the critical protection against tumour development. The virus does this by tagging the protein with a signal that tells the cell to break it down. So, by the most effective means possible, the virus removes the guardian of the genome.

Hepatitis B virus, responsible for the majority of liver cancers that claim one million lives a year, also makes at least one protein that perturbs normal cell signals to promote cancer. Hepatitis C virus, the most common blood-borne infection in the USA, is the leading cause of liver cancer.

Burkitt's lymphoma, identified by the Irish surgeon Denis Burkitt in equatorial Africa, where it is the most common childhood cancer, arises in a type of white blood cell essential to our immune system, and commonly involves the jaw or other facial bones. Anthony Epstein, Yvonne Barr and Bert Achong at the Middlesex Hospital analysed samples collected by Burkitt to reveal that the disease was due to a member of the herpes virus family. This became known as Epstein–Barr virus, the genome of which encodes a number of proteins that can overcome the normal growth restrictions on cells, which is thought to reflect the way it promotes the human lymphoma. It is also the cause of glandular fever.

In general, viruses and bacteria do not directly cause mutations, but they produce equivalent effects by interacting with proteins in cell signalling pathways.

RNA Viruses

Retroviruses have played an important role in the cancer story as the source of the first oncogenes. Moreover, the enzyme they make to convert their RNA genome into DNA for incorporation into the genome of infected cells, reverse transcriptase, has become an essential part of the molecular biology toolbox. However, despite retroviruses being able to cause tumours in animals, particularly the feline and bovine leukaemia viruses, they are rarely associated with cancers in humans – although they can cause other types of fatal disease, as illustrated by the coronavirus epidemics of 2003 and 2020.

Two members of the human T cell lymphotropic virus (HTLV) family are exceptions in that HTLV-1 causes adult T cell leukaemia (an aggressive form of non-Hodgkin's lymphoma) and human immunodeficiency virus (HIV) destroys cells of the immune system, rendering victims susceptible to infection. When infection is by the Kaposi's sarcoma-associated herpes virus, the otherwise rare Kaposi's sarcoma (which is not really a sarcoma because it starts in the lymphatic system) slowly develops.

The Pan-Cancer Project

The technical revolution that followed the Human Genome Project has given birth to something called The Cancer Genome Atlas (TCGA). Started in 2006, the aim of TCGA was to provide a genetic database for three cancer types: lung, ovarian and brain (glioblastoma). Such was its success that it soon expanded to a vast, comprehensive data set of more than 11,000 cases across 33 tumour types, describing the variety of molecular changes that drive the cancers. The upshot is now called the Pan-Cancer Atlas – PanCan Atlas, for short – that we mentioned as an introduction in Chapter 4. The first set of results, released in 2018, represents such a staggering amount of data that you'd need over 530,000 DVDs to store it!

The 33 different tumour types included all the common cancers (breast, bowel, lung, prostate, etc.) and 10 rare types for which the variety of information accumulated included mutations that affect genes, RNA and protein expression, duplication or deletion of stretches of DNA (i.e., copy number variation) and epigenetic changes (which we'll come to in Chapter 9),

The latest instalment of the Pan-Cancer Project, in February 2020, described data from 38 tumour types and reminded us, yet again, of nature's capacity to surprise. The first finding was that, on average, cancer genomes contained four or five driver mutations from a combination of coding and non-coding regions. That's roughly consistent with the accepted estimate over the last few decades. What was unexpected, however, was that in around 5 per cent of cases no drivers were identified, suggesting that there are more of these mutations to be discovered. Another surprise was that chromothripsis, the single catastrophic event producing simultaneously many variants in DNA, is frequently an early event in tumour evolution.

The gene coding regions within chromosomes are protected from damage by regions of repetitive nucleotide sequences called telomeres that 'cap' each end. In normal cells these gradually shorten with age, but cancer cells reactivate the enzyme that makes telomeres. The Pan-Cancer Project analyses revealed several mechanisms by which cancer cells are protected from telomere attrition so that they can replicate indefinitely, and variants transmitted in the germline can affect subsequently acquired patterns of somatic mutation.

Playing Games

Dissecting the mayhem of the cancer cell genome is clearly necessary as a foundation for the rational pursuit of therapeutic drugs, but simply annotating mutations is only the first step. Several hundred driver mutations have been identified, with yet more to come. A figure of about half a dozen required drivers means that the number of critical combinations that can arise is essentially infinite. This raises the problem of identifying combinations, recently tackled by Giulio Caravagna, Andrea Sottoriva and colleagues at the Institute of Cancer Research, London and the University of Edinburgh. They developed an idea that goes back to the 1950s, when a clever chap from Kansas by the name of Arthur Samuel came up with a program for IBM's first commercial computer so that it could play draughts (or checkers as our American friends call it) in its spare time. The program defined the patterns that could be formed by the pieces on the chequerboard so that, given enough of these, the IBM 701 could indicate the optimal moves. Samuel called this *machine learning*, a precursor of the idea of artificial intelligence.

Perhaps the most famous moment in this saga came in 1997 when a later IBM computer, *Deep Blue*, beat the then world chess champion, Garry Kasparov. Unsurprisingly, Kasparov was a bit miffed and accused IBM of cheating – to wit, getting a human to tell the machine what to do. Let's hope that in the end he came to terms with the fact that *Deep Blue* could crank through 200 million positions per second and, however many games Grandmasters have in their heads, they can't compete with that.

The cancer team realized that the mutations driving the evolution of cancer cells emerge as patterns in the sequence of DNA as a cell moves towards becoming independent of normal controls. Think of each cancer as a family tree of mutations, the key question being which branch leads to the most potent combination.

To pick out these patterns they applied a machine learning approach, known as transfer learning, to the DNA sequences from a large number of cancers. They called this 'repeated evolution in cancer' – REVOLVER – aimed at picking out mutation patterns at the heart of cancer that foreshadow future genetic changes and can be used to predict how they will evolve. The method involves sampling different regions of tumours like those we showed in the evolutionary tree of cancer in Chapter 1

(Figure 1.1). Their DNA sequences will have mutations in many genes and there will be overlap between patterns – for example: A, B, C and D / A, D, E and F / D, F, G and H, etc. Using transfer learning, REVOLVER compares sequence data from many patients to reveal underlying patterns of driver mutations that permit categorization of patient sub-groups. It's been applied to sequences from lung, breast, kidney and bowel cancers, but there's no reason it shouldn't work with other tumours. The big attraction is that if these mini-sequence mutation patterns can be associated generally with how a given tumour develops, they should help to inform treatment options and predict survival.

We have in the past referred to the ways cancers evolve as 'genetic roulette' – so perhaps it's appropriate if game-playing computer programs turn out to be useful in teasing out behavioural clues.

The Genomic Cancer Message

We should acknowledge the mind-boggling effort and organization of the Pan-Cancer Project in collecting thousands of paired samples, sequencing them and analysing the output. However, the value of these massive projects is beginning to emerge – and the news is mixed.

One critical trend is that genomic analysis is redefining the way cancers are classified. Traditionally they have been grouped on the basis of the tissue of origin (breast, bowel, etc.), but this will gradually be replaced by genetic grouping, reflecting the fact that seemingly unrelated cancers can be driven by common pathways.

Perhaps the most encouraging thing to come out of the genetic changes driving these tumours is that for about half of them potential treatments are already available. That's quite a surprise, but it doesn't mean that hitting those targets will actually work as anti-cancer strategies – an issue we'll confront in Chapter 8. Nevertheless, it's a cheering point that the output of this phenomenal project may serve as a launch pad for real benefit in the not too distant future.

In this chapter we've looked at the extraordinary range of mishaps that can befall our genetic material to bring about changes in genes that can act as cancer drivers. We've noted the progress towards a genetic basis of cancer classification and we will consider in more detail the situation with regard to chemotherapy when we come to Chapter 8. Before that, we need to look at possible causes of DNA damage and what, if anything, we can do about them.

7 Causes of Cancer That Can be Controlled

We began our cancer odyssey with perhaps the most frequently asked question, 'What causes cancer?', and the shortest reply: 'Mutations'. But that's a biologist's answer. What we really want to know is 'How?' How do these changes come about and, of course, what can we do about them? Broadly speaking, two categories have been long-recognized as the underlying causes of cancer – 'hereditary' and 'environmental'. The former refers to the state of our DNA when we get it – mutations passed to us at birth kick off about 10 per cent of all cancers. The second group includes everything else and in doing so lumps together things that we can't control (e.g., radiation from the ground) with those we can (e.g., smoking). The latter really should include a sub-group: 'self-destruction' perhaps.

And Another Thing

Lurking in the wings for many years has been a potential third cause that arises from a slightly tricky concept – namely the fact that our DNA, the genetic rock upon which all life is built, isn't rock-like at all. In fact, the chemistry of DNA makes it inherently *un*stable. Thinking about it from the viewpoint of evolution, *of course* it's unstable: it is able to permit change as new genes, and hence new proteins, are made and unmade – allowing life forms to evolve. Think of it like close relationships: we're fond of calling such things 'permanent', 'unchanging', 'solid as a rock', even. But they're not: they change all the time, adapting to our shortcomings and to how individuals develop and mature, with the odd bit of chromothripsis thrown in.

With that in mind, maybe it's less surprising to find that DNA reacts with a wide range of chemicals, some that we consume but others arising from the natural reactions of the body – products of metabolism in fact. And then, speaking of shortcomings, there's the truism that 'nobody's perfect' and the realization that this applies to the mechanics of DNA replication as well as everything else. In other words, every time we make a new cell its DNA differs from the original. Cells have remarkably effective methods for correcting most mistakes made during replication but inevitably some get through and become fixed in the new genome.

Although 'replicative mutations' have been known for a while, nobody had come up with a way of measuring how much they contribute to cancers until Bert Vogelstein and Cristian Tomasetti at Johns Hopkins University had the idea of looking at 'stem cells' – cells that can divide to make more of themselves or turn themselves into specialized cell types. They reasoned – bearing in mind that with every division there's a risk of a cancer-causing mutation in a daughter cell – that if you knew the number of stem cells in an organ and you could estimate the total number of divisions over a lifetime, that might relate to cancer risk.

Indeed it did, because it turned out to account for two-thirds of all cancers. In other words, the majority of cancers arise because of cumulative mutations caused by internal agents. Specifically, they found that in lung cancers, 90 per cent of which are caused by smoking and are therefore preventable, heredity plays no significant role and they estimated that 35 per cent of all driver (i.e., cancer-promoting) mutations are due to replication errors. For prostate cancers there is no evidence that environmental factors are significant and hereditary factors account for 5–9 per cent of cases. The remaining 90 per cent or so of driver gene mutations are estimated to be replication errors. Overall, in 18 types of UK female cancer (brain, bladder, breast, cervical, colorectal, oesophagus, head and neck, kidney, leukaemia, liver, lung, melanoma, non-Hodgkin lymphoma, ovarian, pancreas, stomach, thyroid and uterus), they found that in none are inherited mutations statistically significant (note that Paul Broca's findings related to fewer than 10 per cent of breast cancers – the proportion we now know to be caused by abnormal genes passed from parent to child). For almost all these cancers, replication errors were the major source of driver mutations.

Controversial or What . . . ?

It's fair to say that the estimate of two-thirds of all cancers being down to faults in our biochemistry was a surprise to many. It has to be said that there's a continuing debate about the precise numbers – not least because figures for cell divisions in some tissues aren't available and also because of somewhat vaguer problems, such as the extent to which external assaults contribute to replication errors.

Nevertheless, it now seems clear that what Tomasetti and Vogelstein call 'bad luck' can be blamed for a significant number of cancers. That's good because knowing that it's not your fault may help some patients but we need to be wary of promoting that message too strongly in the media.

The upshot of this is that we have to add a third category of 'replication errors' to 'hereditary' and 'environmental' factors. Whatever the proportion that we can put down to bad luck, there are still a great many cancers that can be prevented, and to see how we'll look next at the major factors involved.

Alcohol and Cancer

With that background we'll now turn to the causes of cancer over which we have control – although when we come to obesity we will meet our inner microbial world that in many ways acts as an independent entity. Where better to begin than with alcohol?

Mankind has been brewing and drinking alcohol since at least the Stone Age. Making it involves fermenting sugars from fruits or cereals to produce ethanol (commonly called alcohol) with the release of carbon dioxide. It's the drinking part that causes problems because 20 per cent is absorbed by the stomach and the small intestine, from which it passes into the bloodstream *en route* to all tissues and organs in the body, apart from fat tissue in which it doesn't dissolve.

Most people learn at an early age what the most obvious and immediate effects of drinking alcohol are, although, truth to tell, we still don't really understand how it works. Chronic exposure to alcohol changes the way cells work, especially those of the liver, which is why cirrhosis is a major long-term effect of drinking. In the short term it can, of course, go straight to our heads – literally so by crossing the blood–brain barrier and interacting directly with brain cells. It's also an irrefutable fact that women, being on average smaller than men, are more affected by drinking the same amount of alcohol. It's an effect of concentration (i.e., dilution) though the result may be a loss of concentration of the mental variety, if nothing else.

To make matters worse, nature has dealt women a rather unhelpful hand. The enzyme alcohol dehydrogenase (ADH) breaks down alcohol, getting to work while it is still in the stomach. Remarkably, humans have seven different kinds of ADH, but women, especially when they're young, make less ADH than men, with the result that proportionally more alcohol enters the bloodstream from the stomach.

Circulating alcohol can affect all tissues because it partitions into cell membranes, where it perturbs signalling – an effect that is particularly pronounced in the brain. Alcohol exerts its effects because it's really a general anaesthetic (ether, one of the first anaesthetics used in medicine is, in effect, two molecules of alcohol joined together) – hence the advice not to drink alcohol within two days before having a general anaesthetic.

Regardless of how it works, there is a clear link between alcohol consumption and an increased risk of breast cancer. In developed countries about 4 per cent of breast cancers are attributable to alcohol, the risk increasing with the amount consumed. One plausible explanation of its effects on the breast is that it makes cells produce more oestrogen, which in turn increases cell growth. Compared with non-drinkers, women who consume one alcoholic drink a day have a very small increase in risk. Those who have 2–5 drinks daily have about 1.5 times the risk of women who drink no alcohol. The American Cancer Society recommends that women limit their consumption of alcohol to no more than one drink per day, advice worth following as excessive alcohol use is also known to increase the risk of cancer of the mouth, throat, oesophagus, bowel and liver. From the point of view of cancer, therefore, the best advice is to be teetotal, but we should note the substantial evidence that low levels of alcohol consumption protect against cardiovascular disease and increase longevity. Alcohol therefore differs from smoking, to which we'll come shortly, in that limited exposure may be beneficial.

Diet

It is arguable that over the last 30 years no aspect of human health has been so commercially profitable as the question of what we should eat, particularly the sub-category of how that impacts cancer. It seems that scarcely a week has passed without some new eating plan being launched or the trumpeting of another piece of expert guidance – all nominally aimed at making us healthier, particularly in the interlinked fields of cancer and obesity. Over that period adult obesity in England has more or less doubled (to around 13 million: 29 per cent) and in the USA has gone from 30 per cent to over 42 per cent – indicating that, however profitable the diet fetish has been for its purveyors, it has failed to hit the important target. It is true that cancer mortality, as we noted in Chapter 3, has declined in both countries since the early 1990s (the last decade saw a drop of 9 per cent in the UK), but it's unlikely that eating habits have played a significant part.

There is no question that diet is important in cancer because the numbers overwhelmingly show it to be a major factor in worldwide cancer variation. The Continuous Update Project provides an ongoing survey of research on nutrition and cancer and its current view is that diet, weight and physical activity level affect the risk of getting 16 types of cancer, including colorectal, breast, mouth, pharynx, larynx, oesophageal, stomach, bladder, kidney, liver, gallbladder, ovarian, prostate, endometrial, pancreatic and lung cancers – in other words, most of them. Their summary is that, after not smoking, the most critical thing in avoiding cancer is being a healthy weight. The American Heart Association has similarly sane and simple advice: 'eat a balanced diet and do enough exercise to match the number of calories you take in'.

We should note that new methods are being applied to this problem, most notably metabolomics, which is the study of metabolites – small molecules involved in metabolism. The metabolome is the complete set of metabolites in a cell, tissue, organ or animal at any time. The method involves extracting metabolites from a sample (e.g., saliva or a piece of tissue) and using spectroscopy (nuclear magnetic resonance or gas/liquid chromatography coupled to mass spectrometry) to quantitate each component. In principle, metabolomics can identify dietary chemical fingerprints and link these to cancer risk, prognosis and survival, but this sophisticated approach is still in its infancy.

It would be tedious to list all the diets that have been promoted. Suffice to say that you can divide defined diets into two broad groups: the unhealthy (often called 'Western' diets) and the healthy ('prudent' diets). The pattern of Western diets generally includes red and processed meats, sugary beverages, refined carbohydrates and salty snacks, whereas the prudent or healthy diet is usually heavily weighted towards vegetables and fruits.

Numerous studies have associated dietary patterns with increased or decreased risk of various types of cancer, while others have found no connection, reflecting the difficulties of determining cause and effect in such a nebulous field. Western diets have been consistently associated with increased bowel cancer risk, while their link to breast, prostate and pancreatic cancers tends to depend on study design (case–control studies indicating a positive association but cohort studies reporting no consistent associations – see Box 7.1). On the other hand, prudent diets have been associated with reduced risk of breast, bowel and lung cancers, with more inconsistent results for pancreatic and prostate cancers.

Box 7.1 Retrospective Studies and Randomized Controlled Trials

There are really two ways to answer the question of whether what we do to ourselves, and particularly what we eat, affects our cancer risk. The first is what epidemiologists call a retrospective study: take a group of people who have (a specific) cancer and select a second group who are as similar as possible except that they don't have cancer. Then ask them what they have eaten for the last 20 or 30 years. These are sometimes referred to as observational studies, and the behaviour of each subject is self-selected – they ate what they fancied, which is good because the investigator hasn't influenced what happened. But you will already have spotted a bit of a pitfall: no matter how much statistical analysis you bring to bear, you can't make the two groups really comparable. Equally problematical is that the accuracy of answers to the question 'What have you been eating?' probably exceeds only marginally that of the answers to surveys concluding that everyone engages in sexual congress 3.5 times per week.

The second type of study is prospective: this is an experiment, sometimes called a randomized controlled trial (RCT), in which the participants are divided into two or more groups and each group is given specific behavioural instructions (in this case, told what to eat) by the investigator and everyone waits to see what happens. This may take some time when dealing with cancer, but the main advantage of the method is that random distribution into the groups reduces bias. One of the biggest problems, of course, is that however hard they try, individuals will vary in how well they stick to the dietary rules. This method has been extensively used to find out whether agents in foods that have been thrown up, so to speak, by observational studies actually do confer protection against cancers.

Though easy to describe, such studies are complicated to design: you have to enrol an awful lot of people before the results mean anything and they take a long time and are therefore very expensive. It might be added that, in the diet field, for the most part they have also been singularly uninformative, the clearest conclusion being that linking specific things we eat to cancer is a desperately tricky business. The guy who ate 25,000 Big Macs makes the point: the worst of his problem is probably not the polar bear's weight of fat he's eaten, but the rather vague things in fruit and veg he's not had time to squeeze in.

Case–control studies are retrospective. They define two groups, one with the disease and one without, and then look back to assess whether there is a statistically significant difference in the rates of exposure to a defined risk factor between the groups. Cohort studies can be retrospective or prospective. The latter follows participants for a defined period to assess the proportion that develop disease. Case–control and cohort studies are not the same as RCTs in which participants are randomly assigned to test groups.

The Nurses' Health Study

One of the biggest prospective studies yet undertaken of diet and health is the Nurses' Health Study, which illustrates the scale needed for such projects. It started in the USA in 1976 to investigate risk factors for cancer and other diseases in women, and to date there have been over 275,000 participants. Among its major findings so far are the following:

1. Lack of physical activity correlates with an increased risk of type 2 diabetes. Conversely, exercise helps breast cancer survival and decreases the risk of the development of heart disease.
2. Obesity increases the risk of type 2 diabetes, heart disease, breast and pancreatic cancers, psoriasis, multiple sclerosis, gallstones and eye disease.
3. Oral contraceptives decrease the risk of ovarian cancer and have no significant effect on breast cancer.
4. Postmenopausal hormone therapy lowers the risk of heart disease while hormone combinations (progesterone and oestrogen) associate with higher risk of breast cancer.
5. Smoking promotes type 2 diabetes, cardiovascular disease, bowel and pancreatic cancer, psoriasis and multiple sclerosis.

The study also reported that the risk of bowel cancer is increased by eating a lot of red meat and reduced by taking folic acid in multivitamin supplements. Breast cancer incidence is unaffected by fat and fibre intake, but is increased by one-third in response to moderate amounts of alcohol (1–3 drinks per week), even though the latter makes you less likely to have a heart attack (unless, of course, you read the cancer predictions!).

One other point is that a diet rich in vegetables and fruits can lower blood pressure, reduce the risk of heart disease and stroke and of eye and digestive problems, prevent some types of cancer and have a positive effect upon blood sugar, which can help keep appetite in check.

Meat and Vegetables

The difficulty with food is that it can seem as though you can't win. A lot of things we eat contain DNA-damaging agents or lack an essential nutrient. The advice to 'eat your greens' is generally sound: cabbage, for example, has lots of vitamins C and K, but it also contains chemicals called goitrogens that can interfere with the function of the thyroid gland, which is why those with thyroid problems are advised not to eat it. Forty years ago, Richard Doll and Richard Peto came up with the estimate that diet may contribute to one-third of all cancers. Since then, factors like the replicative mutations we mentioned above suggest that 20 per cent might be a more accurate figure but, whatever the number, there remains the problem that there is very little *direct* evidence linking foodstuffs to cancer development.

Antioxidants illustrate the problem. These are chemicals that can neutralize free radicals – highly reactive chemicals formed when atoms or molecules gain or lose electrons – that are produced as by-products of metabolism that can damage DNA. Sometimes called 'free radical scavengers', beta-carotene and vitamins A, C and E (alpha-tocopherol) are examples of dietary antioxidants. Despite their popularity as supplements, there have been at least nine RCTs that, collectively, provided no evidence that they were of any use in cancer prevention. Indeed, for smokers beta-carotene supplements may actually increase the risk of lung cancer.

Another prominent player in the diet saga is red meat, eaten by man throughout his history because it looks good, tastes good and gives you lots of protein and iron. However, many large and seemingly well-conducted studies have shown that you're a bit more likely to get bowel and stomach cancers if you eat lots of red or processed meat.

The upshot is that we are now advised by the World Cancer Research Fund and the American Institute for Cancer Research to consume less than 500 grams per week of red meat and that 'very little if any' of that should be processed – that is, smoked or otherwise cured. The redness in meat comes from blood, specifically the iron-containing haem group in red blood cells that carries oxygen. Haem is broken down in our gut to give substances that introduce mutations into DNA. In addition, suppliers often add chemicals to meats to give colour and flavour and to stop bugs growing, and the evidence from animal studies is that these too can damage DNA, thus helping to promote cancer. In addition to bowel and stomach cancers there are several reports of a strong association between meat eating and lung cancer. However, to underline how complex this field is, there are also large and evidently well-conducted studies showing no link between meat eating and cancers. A further possibility, consistent with the global epidemiological pattern of colorectal cancer incidence, is that potentially oncogenic bovine viruses may be present in beef as contaminants. These could cause latent infections in the colorectal tract, exerting combined effects with chemical carcinogens produced during cooking. And you can't opt out by becoming a trendy veggie because they too have an increased bowel cancer risk, according to some studies.

Despite that warning, the desirability of eating lots of fruit and vegetables has been a health mantra for many years, although we have little idea of why they are so good for us. However, in a small step forwards, Elizabeth Sattely and her colleagues at Stanford University in California have shown that brassicas (including broccoli, cabbage and cauliflower) make glucosinolates (chemicals that confer their flavour) and that bacteria in the gut transform these into isothiocyanates – and these have protective effects against certain cancers. They tracked down the microbe responsible (*Bacteroides thetaiotaomicron* – one of the most common bacteria in the human gut) and identified the set of genes it uses to carry out this transformation.

Folate

The several naturally occurring forms of folate are vitamins involved in the synthesis of DNA. A synthetic version, folic acid, is a popular food supplement and several studies, both retrospective and prospective, suggest that it protects against cancer, consistent with its importance in maintaining the integrity of DNA. With that in mind, you might reason that being low on folate favours cancer, and indeed several studies have shown that blood folate levels have an inverse association with bowel cancer – that is, the more folate you have the better. There is a clear example of benefit from folic acid supplementation in the 75 per cent decrease in spina bifida that results from daily intake prior to conception.

However, things are never simple in the diet field and there are reports that diet supplementation can actually increase cancer risk. The conflict arises because folate is required to maintain DNA, but in excess it may block DNA repair, thereby promoting cancer.

It may be, therefore, that diet supplementation with folate protects against bowel cancer in people with low levels of circulating folate but might be a serious risk for individuals with higher, more natural levels. All of which indicates the hazards of advocating diet supplementation for general populations without considering the biochemistry of the individual.

Calcium

A diet rich in calcium is well known to be a good thing because not only does it give you strong bones but, according to several prospective studies, you are less likely to get bowel cancer. Against that are studies, including one of 36,000 cases, concluding that consuming extra calcium has no effect. As with folate, the available data scarcely make the case for calcium supplementation and, although it is popular, raising calcium intake will not benefit those on healthy diets whose levels are normal.

Fibre

Dietary fibre or roughage is stuff we eat but can't digest. It does an important job in taking up water and generally helping our insides work. Many surveys have led to the well-publicized advice that eating plenty of fibre helps to prevent bowel cancer. Notable among these is the European Prospective Investigation into Cancer and Nutrition (EPIC) study, showing that 35 grams per day of fibre reduces the risk by 40 per cent compared with 15 grams per day.

This was a particularly powerful study because it involved over half a million (520,000) people from 10 European countries. Almost inevitably, there are other studies (e.g., the Polyp Prevention Trial and the Wheat Bran Fiber Trial) that show no protective effect, and at least one concluding that a relatively high-fibre diet protects men better than women. Overall, the fibre surveys are another illustration of the pitfalls lurking in the diet and cancer field, not least the variation in study duration and the fact that follow-up periods are short relative to the many years over which cancers usually develop.

With all this uncertainty, we might take a moment to salute Stephen O'Keefe and colleagues from the University of Pittsburgh and Imperial College London for coming up with a simple experiment and some pretty astonishing results, although we should reveal at the outset that they confirm that a high-fibre diet can substantially reduce the risk of colon cancer. The experiment compared what happened to two groups of 20, one of African Americans and the other from rural South Africa, when they swapped diets for two weeks. The design was indeed simple, but the experiment involved a huge amount of work. The Western diet was, of course, high-protein, high-fat and low-fibre, whereas the typical African diet was high-fibre, low-fat and low-protein. Just to be clear, the American diet included beef sausage and pancakes for breakfast, burger and chips for lunch, etc. The traditional African diet comprises corn-based products, vegetables, fruit and pulses (e.g., corn fritters, spinach and red pepper for breakfast).

Almost incredibly, within the two-week duration of these experiments there were significant, reciprocal changes in both markers for cancer development and in the bug army – the microbiota – inhabiting the digestive tracts of the volunteers. The dreaded colonoscopy revealed polyps (tumour precursors) in nine Americans (that were removed) but none in the Africans. Cells sampled from bowel linings had significantly higher proliferation rates (a biomarker of cancer risk) in the African Americans than in the Africans. After the diet switch the proliferation rates flipped, decreasing in African Americans, while the Africans now had rates even higher than in the starting African American group. These changes were paralleled by an influx of inflammation-associated cells (white blood cells) in the Africans who were now on a high-fat diet, while these decreased in the Americans on their new, high-fibre diet.

Equally amazing, these reciprocal shifts were also associated with corresponding changes in specific microbes and their metabolites. It really is quite remarkable that these indicators of cancer risk manifested themselves so rapidly following a change to a typical Western diet. Of course, 'markers' are one thing, cancer is another. As one of the authors, Jeremy Nicholson of Imperial College London, said: 'We can't definitively tell from these measurements that the change in their diet would have led to more cancer in the African group or less in the American group, but there is good evidence from other studies that the changes we observed are signs of cancer risk.'

Sugar

We have mentioned the prominence of sugar in the Western diet, and the adverse effects of its over-consumption have been well publicized, leading to calls for draconian measures to limit how much of it we eat.

We should, therefore, summarize the position. Sugar consumption worldwide has gone up three-fold in the last 50 years. According to the American Heart Association, adults in the USA now consume an average of 77 grams of sugar per day. In the UK Public Health England has reported that consumption is also three times the recommended amount and rising, despite the effects of a sugar tax and efforts to persuade food producers to reduce sugar in their products to aid the fight against obesity.

The problem is, of course, that sugar is a great source of calories and that the more calories you shovel down – in whatever form – the bigger you tend to become. And, of course, most of us like the taste.

The World Health Organization (WHO) recommendation is that we should obtain 5 per cent of our daily calories from sugar, but sticking to that limit is more difficult than it appears, not least because, as the WHO points out: 'Much of the sugars consumed today are 'hidden' in processed foods that are not usually seen as sweets. For example, one tablespoon of ketchup contains around 4g (about one teaspoon) of free sugars. A single can of sugar-sweetened soft drink contains up to 40g (about 10 teaspoons) of free sugars.'

Very roughly an 'average' person needs about 2,100 calories a day and 160 grams of sugar would give between one-third and one-quarter of that total requirement. Taking into account the problem of 'hidden' sugar, the American Heart Association recommends a daily sugar intake of no more than 36 grams for men and 25 grams for women but the consumption figures we've just noted suggest that their advice is falling on deaf ears. For a historical perspective the recommendation of 36 grams is about three times as much sugar as the denizens of Great Britain were allowed during the Second World War under rationing – a period when our diet is generally considered to have made us healthier than we've ever been. So you could say an element of control has been lost.

The '2,100 calories' above are 'food calories', the unit sometimes used in nutritional contexts. It's 1000 times greater than 'scientific' calories, or gram calories (cal). Scientifically, therefore, we mean 2,100 kilocalories (kcal) – which is why your fruit juice carton may tell you one glass contains 50 kcal. And, just to stop you wondering, 1 calorie is the heat (energy) you need to raise the temperature of 1 gram of water from 14.5 °C to 15.5 °C.

Sugar consumption has sky-rocketed, eating too much of it unbalances your diet and bad eating habits can cause obesity and metabolic syndrome. But these things aren't black and white: 20 per cent of obese people have normal metabolism and a normal lifespan, while 40 per cent of those of normal weight will get metabolic syndrome diseases. So, while obesity indicates metabolic abnormality, it is not *per se* the cause.

The underlying science remains a matter of debate. What is not in question is that we eat more sugar than we need, and the real crunch is that sugar is like tobacco and alcohol – no, it doesn't make you smelly or do Sinatra impressions – but it is addictive. It actually manipulates your pathetic brain cells so you keep asking for more. So we're seduced into eating more and more of something that can help us get fat and ill. What's to be done? Lenin, who was fond of asking this question, would have dealt with it in a trice by limiting sugar supplies to one-tenth of the dietary minimum and seeing who survived. Ah! The good old days.

In the UK we have at least started to tackle the problem by taxation, but that has failed to stem the rise in consumption, let alone reverse it. This is unsurprising, given that to have an effect on sugar you'd need a huge price increase across a vast range of foods – fruit juice, 'sports' drinks, chocolates, sweets, cakes. Which rather suggests that the only solution is a return to rationing for all foods.

Best Practice

Making sense of either epidemiological surveys of diet or clinical trials of drugs can be tough. We've already noted how difficult it is to study small associations when there is a huge number of unaccountable variables. It's sometimes hard to discern whether these have been adequately accounted for in a given study or not – in other words, whether we should consider it 'seemingly well-conducted research' or throw it out.

It's a regrettable fact that in the diet field you also have to ask who is performing the study and whether they have an ulterior motive, your suspicion being founded on the fact that Americans alone spend in the region of $40 billion annually on dietary supplements (in the UK it's over £400 million), vigorously promoted as beneficial to health but mostly doing nothing at all. Predictably, given the sums involved, the art of self-promotion for profit on the back of pseudoscience is widely practised on television, in the media and in numerous advertisements telling you that a supplement is 'scientifically proven'.

Fortunately there are islands of sanity in this commercial swamp, notably Gary Taubes, who has written extensively on the hazards of sugar, and Ben Goldacre, who has debunked both Big Pharma and individuals with gusto. Furthermore, there is the Cochrane Collaboration (see Box 7.2), a British international charitable organization formed in 1993 to sort medical research findings so that it's easier for non-experts to make informed choices about health interventions. To do this they provide updated, clear summaries of published work in plain English. So let's turn to the Cochrane Collaboration to

---

Box 7.2 The Cochrane Collaboration

You may by now have the impression that the only thing more tricky than carrying out clinical trials is making head or tail of the results. However, help is at hand in the shape of the Cochrane Collaboration (www.cochrane.org), a non-profit consortium dedicated to analysing and summarizing the literature on healthcare interventions. It has literally been a lifesaver by providing doctors with comparative analyses of trials. Its other great virtue is that it is set up to inform non-scientists as well by giving summaries in 'plain English'. These explain, very briefly but clearly, why the question is being asked, how the trials were set up, the main results and the conclusions. For dietary calcium, one of the most recent reports concludes that although there is evidence that calcium supplementation might make a modest contribution to the prevention of bowel cancer, there is not sufficient evidence to recommend its general use. A corresponding report on dietary fibre concludes that there is currently no evidence to suggest that increased intake will reduce the incidence or recurrence of bowel cancer within a two- to four-year period.

see what they make of the vast mass of data that has accumulated on the subject of diet:

> Don't eat much red and processed meat, limit alcohol consumption, don't over-eat and balance calories consumed with exercise taken, stick to a diet rich in fruits, vegetables and whole grains and low in less-nutritious, refined grain and sugar-rich processed foods.

That's it.

So let's give nature the last word on the vexatious subject of diet by observing that living systems are marvellous, balanced machines – that is, they have evolved a large number of regulatory mechanisms to maintain internal conditions in the face of environmental challenges. Anything we put in is tweaking a finely tuned machine and there's always a lot to be said for leaving well alone – and don't overload it with sugar.

Obesity

Over 650 million people in the world are obese, three times as many as 45 years ago, and nearly two billion adults (18 years and older) are overweight. The body mass index (BMI), a measure of body fatness, is calculated by dividing weight in kilograms by the square of height in metres. Below a BMI of 18.5 kg/m^2 is underweight, the normal range is 18.5–25, overweight is 25–30 and obese is over 30. Almost one in three of our British readers will be obese and another one-third will be overweight. For the USA the American Cancer Society has a spectacular series of maps in which each state is coloured red if more than 55 per cent of adults are overweight. In 1992 not one state was red – *not one.* Now turn to the 2007 map: it looks like *The Empire Returns* (British of course) – *every* state is red. *Every single one, including Hawaii*, and if you think the 49th fell off their map and might help, think again: the overweight figure for Alaska is 66 per cent. The current state is that about 70 per cent of American adults are overweight or obese and 37 per cent are obese.

The basic fact behind these figures is metabolic imbalance. Metabolism refers to all chemical reactions involved in keeping cells and the organism alive. Catabolic reactions break down molecules to obtain energy, while anabolism is the synthesis of compounds needed by the cell. Our bodies expend energy either in these chemical reactions or as heat loss, and that expenditure needs to be matched by the energy we consume as food, no more and no less. A number of conditions – increased blood pressure, high blood sugar, excess body fat around the waist and abnormal cholesterol – result from disruption of this balance. As a group, these conditions are called metabolic syndrome: often accompanied by a pro-inflammatory state, it increases cancer risk and cancer-related mortality.

As with other potential causes of cancer, establishing an obesity link is not straightforward. Nevertheless, there is consistent evidence that higher amounts of body fat are likely to lead to a variety of life-threatening conditions, including diabetes, heart and circulatory disease, sleep apnoea, gallstones, degenerative disease and some cancers. Associated cancers include those of the bowel, kidney, liver, oesophagus, pancreas, endometrium, ovary and breast. For most of these cancers the risk is roughly double that in people of normal weight, although it's rather less in ovarian and bowel cancer (10–30 per cent higher). The risk of breast cancer in women increases after menopause, when oestrogen is produced mainly by fat tissues rather than the ovaries. The risk is 20–40 per cent higher in obese postmenopausal women compared with those of normal weight, possibly because excess fat tissue increases oestrogen levels.

The influence of excess weight on cancer risk varies widely between different cancer types, but a measure of its importance is that in the UK it is the second biggest preventable cause of cancer, responsible for more than 1 in 20 cases. If we all had a BMI of less than 25 there would be 12,000 fewer cancers a year. In the USA, about 28,000 new cases of male cancer (3.5 per cent) and 72,000 female cases (9.5 per cent) arise from excess weight, and overall obesity kills 300,000 Americans each year.

Causes

Obesity is a complex, heritable trait influenced by the interplay of genetic and environmental factors. The most familiar obesity-associated mutation affects the leptin gene that encodes a hormone secreted from fat cells that decreases appetite. However, only a very small number of families have been found who carry leptin mutations and, although other mutations can drive carriers to overeating, they are even rarer. Genome analysis has now unearthed more than 50 genes that show associations with obesity, but the effects appear to be very small.

It is therefore true to say that obesity is not always caused by gluttony or an unfortunate predilection for fast food, but the time trends clearly show that we cannot blame genetics for the obesity epidemic.

Effects

Obesity is associated with abnormal levels of hormones involved in growth (e.g., insulin, oestrogens and leptin) and it's generally thought that their raised levels also favour cell proliferation and tumour growth. Insulin resistance – the failure of muscles, fat and liver to respond adequately to insulin so that blood glucose is not utilized for energy – is promoted by excess fat. This leads to hyperinsulinemia – increased blood levels of insulin and insulin-like growth factor-1. These changes precede the development of type 2 diabetes and contribute to bowel, kidney, prostate and endometrial cancers. Overweight and obese individuals are more likely than those of normal weight to have conditions or disorders that create chronic local inflammation that in time can cause DNA damage and cancer. For example, gastroesophageal reflux disease, more common in obese individuals, causes inflammation that can lead to Barrett's oesophagus (a pre-cancerous condition) and oesophageal cancer.

Nevertheless, despite the figures showing a clear obesity–cancer link, it's been a slow business to unearth molecular connections between obesity and cancer. However, there has been some recent progress. First is the finding that two things happen as obesity develops: the number of fat (adipose) cells goes up but they also grow bigger (i.e., the fat cells themselves are fatter). This causes a knock-on effect that is even more serious: the fat cells attract other cells from the circulation and this cellular cooperative releases signals that can drive tumours. The cells recruited into the tumour can 'talk' directly to the tumour cells, and they do this by releasing the hormone messenger leptin that stops us feeling hungry. These cells are fibroblasts – part of the supportive framework of cells and tissues, so they're 'cancer-associated fibroblasts' – rather than fat cells and several studies have shown that leptin can drive proliferation of tumour cells *in vivo* and *in vitro* and in human cancers.

Another piece of the jigsaw has come from Xavier Michelet and colleagues from institutes in Boston, Kentucky and Ireland, who have revealed that obesity can damage our anti-cancer defences. It does this by taking aim at natural killer cells (NK cells) – a sub-group of white blood cells (lymphocytes) that are a key part of our immune system. NK cells attack tumour cells directly, making holes in their outer membranes and essentially blowing them up. Obese individuals have raised levels of circulating free fatty acids (FFAs), and these are taken up by NK cells. The fatty blobs can be seen under the microscope, and it's no surprise that these fat-laden cells become paralysed. Critically for their anti-tumour activity, this disruption cuts production of the proteins that target tumour cells (perforin and granzymes).

At long last we have a clear molecular link between obesity and cancer: the raised levels of FFAs push a metabolic switch in NK cells that blocks their ability to kill tumour cells – so a major repressor of tumour growth is overcome.

The Obesity Microbiome

In Chapter 6 we mentioned in passing a sobering fact about being human – we're mostly bugs, which is to say that on a cell-to-cell basis the microbes in our bodies outnumber us by about 1.3 to 1. Your microbiota – the 2,000 or so assorted species of bacteria that have made you their home – mostly (99 per cent) reside in your digestive tract. They make up under 3 per cent of our mass so they pass largely unnoticed, but our 'gut flora' are important because they complete the extraction of energy from food and they make some essential vitamins. However, this vast bacterial army, toiling away in the dungeon of our innards, is now being revealed as an important player in our overall health, and in particular in obesity and cancer.

Bugs Tummy

The bacterial army, of course, has its own genomes and the total number of microbial genes can be estimated from faecal samples (i.e., stools). Remarkably, they outnumber our own genes by several hundred times. There are about 20,000 human genes: the bugs muster several million. In the context of obesity, two major sub-groups of the army have emerged – *Bacteroidetes* and *Firmicutes* (Bs and Fs). In obese animals (including humans) there's a startling shift in the proportions of Bs and Fs: the number of Bs is halved while Fs double, compared to normal numbers. Because the gene pool size differs between Bs and Fs, the significance of this swing is that the total of microbial genes in our gut drops dramatically if we become obese:

Fewer genes = more body fat.
More genes (a more diverse microbiome) = healthy status.

Cause or Effect?

Does obesity lower the bacterial gene pool or does its decrease cause obesity? The answer comes from mice born under aseptic conditions that don't have any gut microbes – they're 'germ-free' mice – and as they develop they have less body fat than normal mice. But transfer a sample of gut bacteria from a normal mouse and the 'germ-free' mice double their body fat in a couple of weeks. Do the corresponding experiment with human gut microbes and, if they're from someone who's obese, the mice become likewise, if fed a high-fat rather than a normal diet. These results suggest that microbiomes associated with obesity increase the efficiency of energy extraction from food.

Escherichia Coli and Bowel Cancer

E. coli is probably the most familiar bacterium that is common in human and animal intestines, and is part of our normal gut flora (i.e., bowel bacteria). Hans Clevers and colleagues at the Hubrecht Institute in the Netherlands have recently shown that a toxin called colibactin, produced by one strain of *E. coli*, causes unique patterns of DNA damage in cells that line the gut. They examined 3,600 Dutch samples of various cancer types and found these 'fingerprints' were present much more often in bowel cancers than other cancer types. In more than 2,000 bowel cancer samples from the UK the colibactin fingerprints were present in 4–5 per cent of patients, suggesting that colibactin-producing *E. coli* may contribute to 1 in 20 bowel cancer cases in the UK.

This is the first evidence that a bacterium native to our gut can cause DNA damage that appears to associate with bowel cancer. The challenge now is to establish DNA damage fingerprinting as a way of screening for this risk factor and then finding ways to reduce the presence of toxin-producing bacteria in the gut.

Bugs and Cancer: Drivers or Mirrors?

We've already seen much evidence that where obesity lurks cancer looms. Indeed, transferring microbiota to germ-free mice promotes a wide range of tumours and, conversely, depleting intestinal bacteria slows the growth of liver and bowel cancers. The latter occur more frequently in the large intestine than in the small, which has a much lower microbial density.

In yet another twist to the microbe tale, another species, *Fusobacterium* – related to the *Bs* and *Fs* – has been known for some years to be enriched in human bowel cancers compared to non-cancerous colon tissues, suggesting, though not proving, that *Fusobacteria* may be pro-tumorigenic. In the latest instalment, Susan Bullman and colleagues have shown that not merely is *Fusobacterium* part of the microbiome that colonizes human colon cancers, but that when these growths spread to distant sites (i.e., metastasize) the little *Fs* tag along for the ride!

In other words, when metastasis kicks in it's not just the tumour cells that escape from the primary site, but a whole community of host cells and bugs sets sail on the high seas of the circulatory system. But doesn't that suggest that these bugs might be doing something to help the growth and spread of these tumours? And if so, might that suggest that . . . of course it does, and Bullman and Co. did the experiment. They tried an antibiotic that kills *Fusobacteria* (metronidazole) to see if it had any effect on *F*-bearing tumours. Sure enough, it reduced the number of bugs and slowed the growth of human tumour cells in mice. We're still a long way from a human therapy but it is quite a thought that antibiotics might one day find a place in the cancer drug cabinet.

Debugging

In the next chapter we'll draw a parallel between cancer chemotherapy and bacterial resistance to antibiotics, but, in light of the evidence that gut bugs can determine whether we put on weight, you may already be wondering whether our uninhibited use of antibiotics may have unintended consequences. Beyond the categories of 'narrow spectrum' and 'broad spectrum', antibiotics are not specific and target a range of bacteria – as you discover when you take penicillin for a throat infection and get diarrhoea.

Martin Blaser's group at New York University tackled this question of collateral damage and showed that some antibiotics make mice put on weight. If they are on a high-fat diet they build up even more fat. The antibiotic treatment had, of course, affected the mice microbiota but, astonishingly, the changes happened *before* the mice became fat. Even more extraordinarily, one course of antibiotic treatment imprints these effects on the animal permanently: it acts for life.

It seems incredible that a short drug pulse of the kind given to children to cure ear infections can have permanent effects, albeit in mice. Some gut bacteria survive drug treatment better than others, and the shift in microbiota balance improves the efficiency of digestion, providing more energy.

The evidence is not yet conclusive, but since the 1940s the escalation in antibiotic use has paralleled the obesity explosion – raising the thought that this may not be coincidental. It's food for thought that there are 30 million antibiotic prescriptions a year in the UK. Americans get through over 250 million and American children average 15 courses of antibiotics in their early years.

The Lung Cancer Microbiome

Focus on the microbiome has mainly been in the context of obesity and bowel cancer, but if all this is going on in the intestines you might well ask 'What about the lungs?', because their role of extracting oxygen from inhaled air means they are covered with the largest surface area of mucosal tissue in the body. They are literally an open invitation to passing microorganisms – as we all know from the ease with which we pick up infections. The question 'Could the bug community play a role in lung cancer?' is particularly pressing not only because lung cancer is the major global cause of cancer death, but also because 70 per cent of lung cancer patients have bacterial infections and these markedly influence tumour development and patient survival.

Tyler Jacks and colleagues at the Massachusetts Institute of Technology, using a mouse model for lung cancer (in which two mutated genes, *Kras* and *TP53*, drive tumour formation), found that germ-free mice (or mice treated with antibiotics) were significantly protected from lung cancer in this model system. As lung tumours grow in this model the total bacterial load increases. This abnormal regulation of the local bug community stimulates white blood cells (T cells present in the lung) to make and release small proteins (cytokines, in particular interleukin-17) that signal to neutrophils and tumour cells to promote growth. This really significant finding reveals that cross-talk between the local microbiota and the immune system can drive lung tumour development. The extent of lung tumour growth correlated with the levels of bacteria in the airway but not with those in the intestinal tract – so this is an effect specific to the lung bugs. Indeed, rather than the players prominent in the intestines (Bs and Fs), the most common members of the lung microbiome are *Staphylococcus*, *Streptococcus* and *Lactobacillus*. In a final twist, Jin and colleagues took bacteria from late-stage tumours and inoculated them into the lungs of mice with early tumours – which then grew faster. Thus has been revealed a hitherto unknown role for bacteria in cancer and, of course, the molecular signals identified join the ever-expanding list of potential targets for drug intervention. The scope extends beyond bowel and lung cancers as sequence data from the Cancer Genome Atlas has revealed unique microbial signatures in tissue and blood from most major types of cancer, suggesting the possibility of a microbiome-based approach to diagnosis.

Marching to a Beat

As the importance of our inner army is unveiled, yet another amazing feature has come to light: it keeps time. That is, the amounts of different species oscillate in synchrony with the day/night cycle – in other words, they follow the same circadian rhythm of light-responsive behavioural changes that regulate the sleep–wake cycle over a roughly 24-hour period. These can be large fluctuations – a species can double in amount in six hours and be back to the initial level six hours later.

A familiar example of biological rhythms is what happens when we upset them by flying long distances on an east–west axis. Mice have the same problem: just like us, their clock is disturbed by jet lag. This can be simulated by shifting their ambient light–dark cycle forwards or backwards by eight hours every three days (cheaper than shuttling them business class across the Atlantic). This largely blocks microbiota rhythmicity and reduces the total number of gut bacteria, in turn raising blood sugar level and leading to obesity. This sequence absolutely depends on perturbation of the bacteria because it is replicated in germ-free mice after transfer of 'jet-lagged' faeces. These remarkable results indicate that our symbiotic relationship with microbiota extends to their being governed by the same circadian rhythm, driven by a 'master clock' in the brain that synchronizes all body clocks.

The Mycobiome

Important though they are, bacteria aren't the only members of the microbiome – it also includes viruses, various single-celled parasites (protozoa) and fungi. Fungi are a group of microorganisms, familiar to gardeners worldwide, that includes yeasts and moulds as well as the more familiar mushrooms. There's estimated to be several million species of fungi, although only about 120,000 have been described. Some we can eat, some can kill us and, of course, there's magic mushrooms.

With all this diversity, you might wonder whether any fungi have found their way into us to share the delights of the human body alongside bacterial microbes. Of course they have: most people will have heard of candidiasis – caused by yeasts that belong to the genus *Candida*, normal flora that find a niche in places like the gut, vagina, skin and mouth and usually doesn't give us any trouble – although its overgrowth gives thrush in the mouth. But, truth to tell, we've known very little about fungi in us until recently, when the power of DNA sequencing has started to be applied to the topic. This has confirmed that we do carry lots of fungi around with us, albeit that they are only a tiny fraction of the microbial community (somewhat less than 0.1 per cent).

This fungal force of microbes is known as the mycobiome (as distinct from the microbiome) and, in contrast to bacteria, there was no evidence that it has a role in cancer – until, that is, the 2019 publication from New York University School of Medicine by Berk Aykut and colleagues, showing that fungi travel from the gut to the pancreas where a particular species can actually give cancer a helping hand. The cancer in question is pancreatic ductal adenocarcinoma (PDA), which has a particularly dismal prognosis.

Aykut's group first used DNA sequencing to look for fungus-specific sequences in the pancreas of humans with PDA and in mouse models of PDA. They'd previously shown that the bacterial load goes up by about 1,000-fold in tumours compared with healthy tissue and, lo and behold, they found a similar increase in fungi. Next they tagged strains of fungus with a fluorescent label and showed that the cells could migrate from the gut to the pancreas of mice in under 30 minutes.

They then tracked down a protein called mannose binding lectin (MBL), expression of which is associated with poor survival in human PDA patients. MBL is a 'serum protein', meaning that it floats around in blood. This led to the discovery that MBL can bind to the surface of fungal cells and, when it does so, it changes shape to permit activation of a relay of signal proteins called the complement system. This is part of our immune system, enhancing the capacity of antibodies and phagocytic cells to clear microbes from the circulation. The complement system is pretty amazing because, while it can trigger an immune response against invading pathogens, it can also switch on inflammatory pathways that help cells grow and move around – in other words, give a helping hand to tumours.

One species of fungus seems to be particularly abundant in PDA: the genus *Malassezia*. This is true for both mouse and human tumours, and perhaps that shouldn't surprise us as *Malassezia* is the most abundant fungal species in mammalian skin, accounting for more than 80 per cent of our skin mycobiome.

In a final exciting experiment, Aykut and colleagues showed that antifungal drugs halted PDA progression in mice and improved the ability of chemotherapy to shrink the tumour. This obviously raises the notion that if we can find ways of shifting the balance of fungal communities or interfering with the link to the complement system, we might have a completely new line on desperately needed therapies for this disease.

You could, then, think of this unseen army of tiny cells within us as an organ in its own right. Bacterial composition changes in obesity, it exerts control over some cancers, and if you try to avoid sugar be wary of swapping to saccharin, the most commonly used artificial sweetener, because it causes big shifts in the proportions of different types of gut bacteria. In the diet-switch experiment, changing Americans to a high-fibre diet shifted bacterial balance (from *Bs* to *Fs*). It's a delicately poised occupation upon which we depend for survival – and it's one that we disturb at our peril.

Tea (and Coffee)

We've considered the hazards of alcoholic beverages, but we should comment on the fact that we drug ourselves with other fluids. The most common hot drinks worldwide are tea and coffee, both of which contain caffeine. Like alcohol, caffeine can cross the blood–brain barrier and act directly on the central nervous system – so next time you brew up, bear in mind that you're about to take a psychoactive drug. It's also well known as a diuretic – increasing the volume of urine produced – although regular users get used to this side-effect.

So, is there any evidence that all this drug abuse of our throats drinking tea or coffee gives us cancer? It's been a popular field with over 50 studies of the association between tea consumption and cancer risk published since 2006. You might guess, correctly, that the findings show a fair degree of inconsistency but, where there is a message, drinking tea seems to reduce the risks of colon, breast, ovary, prostate and lung cancers.

The picture that emerges from the most comprehensive meta-analysis (i.e., assessing the results of previous research) of drinking caffeinated coffee is broadly the same. Coffee intake was associated with reduced risk of oral, pharynx, liver, colon, prostate and endometrial cancer and melanoma but with an increased lung cancer risk. The picture for decaffeinated coffee is less clear, but at least it does not appear to increase risk.

As to how these effects could come about, we might note that the leaves of the *Camellia* tea plant are a particularly rich source of antioxidants and, on a perserving basis, coffee provides even more of them. Substantial proportions appear to be absorbed by the body, so it is possible that they may protect DNA from mutation.

Tobacco

Mention of tobacco and smoking was inevitable in our discussion of cancer statistics: it claims the lives of at least six million people worldwide each year and has been linked to at least 17 different types of human cancer. By way of a reminder, in the USA smoking causes 30 per cent of cancers, just ahead of poor diet (25 per cent), and is responsible for 440,000 premature deaths each year. The WHO estimates that 100 million people died in the twentieth century from tobacco-associated diseases. In addition to cancers these included chronic lung disease and cardiovascular diseases. This field focuses almost exclusively on cigarette smoking, but we should note that smokeless tobacco consumption (e.g., by chewing betel quid, a mixture of tobacco, crushed areca nut and other ingredients) is widespread throughout the world and causes devastating damage to the oral cavity with the potential for malignancy. Chewing betel quid is apparently popular in India: its elimination could halve the country's burden of oral cancer, the highest cancer-related mortality among men aged 30–69.

The seminal study that conclusively linked smoking to lung cancer was published in 1950 by Richard Doll and Austin Bradford Hill. It took a while before this resulted in any action, but gradually over the ensuing 50 years the number of UK smokers declined by about half. The trends in lung cancer deaths viewed alongside cigarette consumption are a striking example of cause and effect. Before we smoked we didn't get the disease: as the number of smokers went up, so did lung cancer, and it's only since we've stopped lighting up that the death rate has started to fall (Figure 7.1). For adults over 16 years of age in the UK, smoking prevalence fell from 28 per cent in 1998 to 14 per cent in 2018. Even so, this is hardly a startling drop, and lung cancer still kills 35,000 per year – about 100 people a day – which is perhaps not so surprising given that Cancer Research UK tells us that *every day* rather more than that number of children, 280, take up smoking.

Perhaps the most depressing aspect of the smoking saga is that the first statistical evidence linking cigarette smoking to lung cancer was published

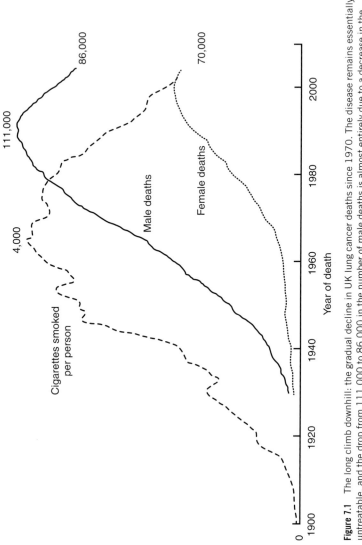

Figure 7.1   The long climb downhill: the gradual decline in UK lung cancer deaths since 1970. The disease remains essentially untreatable, and the drop from 111,000 to 86,000 in the number of male deaths is almost entirely due to a decrease in the number of smokers. The pattern in women is similar but lags behind that of men and has only begun to decline since 2000. The US trends are similar.

not by Doll and Hill in 1950 but 100 years ago in the 1920s. By 1935 the German physician Fritz Lickint felt able to write that there was 'no longer any doubt that tobacco played a significant role in the rise in bronchial cancer' and to coin the term 'passive smoking'. However, because those pre-war studies were carried out in Germany and published in German, they have tended to be ignored.

The gradual decline in smoking in the Western world is, of course, encouraging, but the fact remains that tobacco is killing more than 8 million people each year, set to rise to 10 million by 2030. As ever, the problem is the tobacco companies of the Western world and China that continue to promote cigarette smoking. As we've noted, this is heading towards 2–3 million Chinese killing themselves by smoking in 2030 and it seems certain that tobacco-related deaths in Africa will also rise dramatically.

Of the 8 million a year killed by smoking, about 1.2 million are estimated to die from exposure to second-hand smoke. This problem of involuntary (passive) inhalation, first recognized by Lickint, can nowadays be measured from the blood levels of a substance called cotinine, which is produced from nicotine. The most convincing evidence for second-hand effects is from studies of non-smokers living with smokers, for which pleasure the non-smokers are 20–30 per cent more likely to get lung cancer. Individuals exposed to cigarette smoke at work are at similarly increased risk. This confirmation of Lickint's deductions has eventually led to most countries regulating smoking, with many banning smoking in public places. Bhutan was the first nation to outlaw smoking in all public places. In the UK tobacco advertising is banned, as is smoking in all workplaces. But none of this happened until we'd got well into the twenty-first century! If you're wondering how the country that likes to style itself as the world leader is getting on, Congress has so far managed to avoid passing *any* nationwide smoking ban and left it to individual states – with the result that across the USA laws range from total bans to no regulation of smoking at all!

There is a further problem with second-hand smoke, of course, namely children exposed to it in their home environments, and the subject is complicated yet more by the dawning recognition that 'third-hand smoke' – residual air-borne contamination present long after cigarettes have been extinguished – may also be a significant threat, contributing to a range of diseases, most notably asthma in the children of parents who smoke.

The major target of the chemical cocktail that is cigarette smoke is, of course, the lung, and smokers are indeed about 30 times more likely to get lung cancer than non-smokers. However, the effects of smoking are not confined to the lung: almost all the major cancer types are 2–6 times more likely to develop in smokers than in non-smokers (i.e., mouth and throat, bladder, oesophagus, pancreas, stomach, liver, cervix, kidney and some leukaemias). It is surprising that breast, prostate and endometrial cancers have not, in the main, been associated with smoking – surprising because, for example, chemicals in tobacco smoke cause breast cancer in rodents. There are studies suggesting there may be a link but others that found no convincing evidence, leading the US Surgeon General's 2006 report to comment that the evidence is 'suggestive but not sufficient'.

What is clear, however, is that the age at which women start smoking is highly significant: smoking within five years of the first menstrual cycle almost doubles the risk of breast cancer before menopause. This may be because breast tissue that is still developing is more sensitive to smoke carcinogens.

Smoking and the Genome

The foregoing summarizes the vast amount of epidemiological evidence linking smoking with cancer that has built up over almost 100 years. We've already mentioned (in Chapter 6) the first direct evidence that chemicals in cigarette smoke can cause DNA damage predisposing to cancer – the affected gene being *TP53*. However, it was not until 2016 that the first large-scale study quantified the degree of damage that smoking does to our DNA. Ludmil Alexandrov and colleagues analysed changes in DNA in over 5,000 cancers of types for which tobacco smoking confers an elevated risk. Not unexpectedly, they found that the total mutation burden is elevated in smokers compared to non-smokers in lung adenocarcinoma, larynx, liver and kidney cancers – implying that the chance of acquiring 'driver' mutations is increased. These come in the form of multiple distinct 'mutational signatures' (characteristic combinations of mutation types arising in DNA replication and DNA repair pathways), each contributing to differing extents in different cancers. What was surprising was the absence of mutations in other smoking-associated cancer types, possibly because critical DNA damage was not detected due to being in a small number of clones of cells that expand over time as the symptoms of cancer appear. They estimated the number of mutations in a normal cell due to smoking a pack of cigarettes a day for a year: the highest figure was 150 mutations in the voice box (larynx).

This genomic picture has been enlarged by the finding that within a few weeks of giving up smoking over 100 genes in the nasal airway change expression, reverting from greatly increased to normal levels. This indicates that stopping smoking does lead to at least a partial reversal of lung damage. The genes involved are mainly those switched on by some of the 60 or more carcinogens present in the 7,000 different chemicals in cigarette smoke.

Consistent with this is the observation that the lungs of ex-smokers have a large fraction (20–50 per cent) of cells that appear healthy and have few smoking-associated mutations. It seems, therefore, that stopping smoking leads to the elimination of many DNA-damaged cells and their replacement by normal cells, perhaps derived from a pool of undefined stem cells. How this happens is a mystery, but it's consistent with another major survey conducted by Doll and his colleagues showing that giving up smoking improves the chances of avoiding lung cancer. The earlier the better, of course, but even after the age of 50 stopping reduces risk by over 60 per cent. As Richard Peto, Doll's collaborator, put it 'smoking kills, stopping works'. A last note on smoking is the persuasive evidence that minor variants in a number of genes affect all stages: how likely we are to start, how much we smoke and how easy it is to give up.

In surveying the causes of lung cancer we should note that mesothelioma is caused by the damage that fibres of asbestos do to the delicate membrane sheets that cover internal organs, the lung being especially vulnerable. Here, the initial cause isn't mutation but inflammation that, if sustained, can lead to DNA damage. For asbestos the effect is indeed 'chronic' in that the time lag between exposure and evident disease is typically 20–40 years, which is why the death toll remains high even in countries where the law now reflects its hazards – over 5000 asbestos-related disease deaths in the UK and 12,000–15,000 in the US annually.

8 Causes of Cancer That Are Difficult to Control, Accidents . . . and Other Things

Following the controllables we come to things we either cannot control or are difficult to regulate, together with accidents and some dark episodes that are also part of the cancer story.

Infection

We've already made reference to infection by viruses and via contaminated water and the contribution they make to the global cancer burden. Chronic infection of the kind that can result from drinking impure water eventually weakens the immune system – the body's defence against foreign agents known as pathogens, a group that includes bacteria, fungi, viruses and parasitic worms. It is now recognized that the immune system also recognizes tumour cells and can attack and destroy them too. Inflammation is one of the first responses of the immune system to infection and we'll come back to this important topic when we discuss the tumour environment and immunotherapy. Impairment of the immune response thus attenuates one of the brakes on tumour development.

The bacterium *Helicobacter pylori* gives rise to chronic infection and ulcers in the stomach. However, *H. pylori* can double the risk of stomach (i.e., gastric) cancer and some more potent strains increase the risk by 30-fold. About two-thirds of the world's population carries this bacterium but the vast majority don't get stomach cancer, clearly indicating the involvement of other factors.

The tuberculosis (TB) bacterium (*Mycobacterium tuberculosis*) not only kills two million people a year but is present in latent form in about one in three of us. It's been known for many years that patients with TB have an increased frequency of lung cancer and recent experiments with mice have shown that chronic infection can indeed provoke malignant lung cancer.

There are, of course, numerous other infections that can strike us – one such killed the German physicist Heinrich Hertz by triggering inflammation of his own blood vessels. Hertz was only 36 at the time but he had already proved the existence of electromagnetic waves – we'll come to their involvement in cancer later in this chapter.

Radiation

Everything that we see in the world comes to us quite literally in the form of electromagnetic radiation – Hertz's waves: a stream of particles called photons that carry energy at the speed of light. What we see is only a tiny section of the electromagnetic spectrum – the visible spectrum of wavelengths from about 700 to 400 nanometres (nm). Slightly higher frequencies take us into the UV range (wavelength 400 to 10 nm), where the energy can damage living matter directly. At still higher frequencies, X-rays (wavelengths of 3 to 0.03 nm) have enough energy to pass through solid matter, including the human body. Higher frequencies still bring us to the world of 'ionizing radiation' – a form of high-energy radiation that can release electrons from atoms and molecules, generating ions capable of breaking covalent bonds. Ionizing radiation can therefore directly affect DNA by causing double-strand breaks and by generating reactive oxygen species that comprise both free radical and non-free radical oxygen intermediates such as hydrogen peroxide (H_2O_2), superoxide ($O_2^{\bullet-}$), singlet oxygen ($^1O_2$) and the hydroxyl radical ($^\bullet OH$), which can oxidize proteins and lipids. Collectively, all these changes induce cell death and mitotic failure. Ionizing radiation includes gamma rays, produced by subatomic particle interactions (e.g., in radioactive decay), X-rays that result from high-speed electrons colliding with metal, and alpha and beta particles released by unstable isotopes undergoing radioactive decay. Alpha particles are absorbed by paper and don't have enough energy to pass through the dead, outer layer of our skin. Beta particles (about 8,000 times smaller than an alpha particle) can pass through clothing and skin. Gamma rays and X-rays can pass through all but the densest of materials and travel a long way through air – hence the use of lead and concrete as shields.

Radioactivity is measured in becquerels (Bq), named after Henri Becquerel (1 Bq means that one nucleus decays per second). Radioactivity was formally measured in Curies (Ci), named after Pierre and Marie Curie, but as 1 Ci = 3.7×10^{10} Bq (i.e., an awful lot), the Bq is more convenient. All three were French (albeit Marie was born in Warsaw) and shared the 1903 Nobel Prize in Physics.

Ionizing Radiation

Ionizing radiation can damage living tissues directly, most significantly by creating cancer-promoting mutations in DNA. However, humans have evolved in a background of this kind of radiation from radioactive elements in the earth and in rocks, trace amounts of which are also present in food and water, and in radiation from space. Collectively these sources contribute about 87 per cent of the ionizing radiation we receive. The rest, the avoidable stuff, is artificial radiation of one sort or another, including X-rays used in medicine.

As an example of unavoidable radiation, it's worth spending a moment considering potassium, present in blood and in the cytosol of cells, and one of the most important minerals in the body, helping to regulate fluid balance, muscle contraction and nerve signals. However, it's also the major radio-active emitter in our bodies. There are three naturally occurring forms (i.e., isotopes), two of which are stable, but potassium-40 is radioactive with a half-life of 1.3 billion years – so it's a good job it's only 0.012 per cent of our potassium. We can't avoid it because it's in the earth, though the use of fertilizers increases the amount by several million, million Bq a year. It's therefore present in fruit and vegetables and, because cows will eat grass, a litre of milk comes with about 74 Bq. If you're wondering what you're emitting, the answer is that 5,000 atoms of potassium-40 decay every second in a 70 kg person, releasing beta particles and gamma rays.

Radiation used in medicine comes mainly in the form of X-rays and computed tomography (CT). A conventional X-ray directs the radiation at a part of the body and collects what passes straight through on a photographic film or digital sensor, giving the familiar 2D images. CT also uses X-rays to acquire two-dimensional images, but the radiation beam scans the body and from a large number of such images a 3D picture is pieced together. Invented by Godfrey Hounsfield and Allan McLeod Cormack, this can give an image of whole organs, and it has become an immensely powerful diagnostic tool since its introduction in the early 1970s.

The Swedish physicist Rolf Sievert was a pioneer in measuring the biological effects of radiation and the sievert (Sv) is a measure of the effect of ionizing radiation on the human body. The dose of radiation absorbed is measured in grays (Gy; 1 Gy being the absorption of one joule of radiation energy per kilogram of matter), but the sievert takes into account other factors (type of radiation, exposure time, body part exposed and the volume thereof) to measure the 'relative biological effectiveness'. The dose from a typical chest X-ray is about 0.04 millisieverts (mSv). For a corresponding CT scan the figure is 8 mSv. Our annual dose of 'unavoidable' natural radiation is about 3 mSv, so these typical medical exposures are not a serious cancer hazard – so long as you aren't a baby in the womb.

Radiation is, of course, used in radiotherapy for cancer, the strategy being to target its lethal capacity at tumour cells. This requires much higher exposures than those discussed above, which has driven the development of increasingly sophisticated machines that can target precisely the contours of the tumour with the minimum of collateral damage to healthy tissue.

Abnormal Exposures

The Radium Girls

The first major incident involving accidental exposure to fatal levels of radiation occurred in the inter-war years, when radium was used to make luminous dials for watches and instruments. Young women employed by the US Radium Corporation, who did the radium painting, began to fall ill and indeed die as a result of being told to lick their brushes to maintain their shape. Eventually five 'radium girls' sued the company, finally winning compensation after the case went to the Supreme Court. All of which was somewhat appalling given that radium had been discovered in 1898, the hazards of radium radiation were known and had killed one of its discoverers, Marie Curie, in 1934.

Unethical Human Experimentation in the USA

Copious evidence attests to the fact that, in the twentieth century, numerous unethical experiments were performed in the USA on human test subjects, generally without the knowledge and consent of those individuals. From the 1940s these included human radiation experiments, all under the auspices of the US government. These appalling acts were on a scale that places them in the same category as crimes committed under the German Reich or by the Soviet Union. It was not until 1962 that the FDA were empowered to ban unethical experimentation on humans.

Mercifully we can pass over this dreadful saga but before we leave it we should pay homage to Albert Stevens, a patient being treated for stomach cancer, who in May 1945 was injected with 131 kBq of a mixture of plutonium isotopes, without his knowledge, in a human radiation experiment at the University of California Hospital, San Francisco. Stevens lived a further 20 years before dying of heart disease, thereby surviving the highest known accumulated radiation dose in any human. Because plutonium decays very slowly it remained in his body as he accumulated an effective radiation dose of 64 Sv, an average of 3 Sv per year compared to the annual permitted dose in the USA of 0.05 Sv. The fact that Stevens did not contract cancer, although the radiation he received must have continually damaged his DNA, suggests that his DNA repair systems had been up-regulated sufficiently to compensate.

Hiroshima and Nagasaki

The two atomic bombs dropped on Hiroshima and Nagasaki in 1945 caused an estimated 200,000 deaths mainly from the explosive blasts, subsequent firestorms and from acute radiation poisoning. Extremely high levels of nuclear radiation released in the form of gamma rays and (fast) neutrons killed almost everyone within 1 km of the epicentre. However, the radiation level fell so rapidly that within one week it had returned to normal. Since the war the Radiation Effects Research Foundation has monitored the health of 100,000 survivors, 77,000 of their children and 20,000 people who were not exposed to radiation. This has revealed that cancer rates among survivors were higher than for the unexposed group, with younger people and women being at greater risk. This group had a 10 per cent above normal incidence of solid cancers between 1958 and 1998 and an increased rate of cancers of the blood attributable to radiation but, even so, the number of leukaemias was very small (<4 per cent of the total). The sub-group exposed to a very high radiation dose of 1 Gy (approximately 1,000 times higher than current safety limits for the general public) had a 44 per cent greater risk of cancer over the same time span, but overall even this dose reduced average lifespan by only just over a year. Within the exposed group of survivors there is no significant evidence for mutations caused by radiation being passed to offspring (that is, for the acquisition of germline mutations).

Taken together, the vast amount of data collected over more than 70 years indicates that most survivors did not develop cancer and that if their children had additional health risks they were very small. All of which is somewhat at odds with the general perceptions of the consequences of these bombings and suggests that, although transient exposure to high levels of radiation may cause mutations, the body is quite good at repairing the damage.

Chernobyl

On 26 April 1986 a reactor at the Chernobyl power plant in Ukraine, then part of the USSR, exploded, blowing a large amount of radioactive material into the air. In the worst nuclear accident so far, 1.1×10^{19} Bq was released, some 400 times that of the bombs dropped on Hiroshima and Nagasaki, most of it (60 per cent) falling on Belarus but being subsequently detected across Northern Europe and in North America. The cloud carried over 100 radio-active elements, most decaying rapidly but including iodine-131 (half-life of 8 days), caesium-137 (half-life of 30 years) and strontium-90 (half-life of 29 years) that decay by beta and gamma emissions and thus constitute a health hazard to humans. The thyroid gland is the only one in the body that can absorb iodine, where it's used to make hormones (e.g., thyroxine). Radioactive iodine is given orally to patients with thyroid cancer after surgical removal of the gland to kill any residual tumour cells. In the aftermath of Chernobyl the uncontrolled uptake of iodine-131 caused thyroid tumours to develop in a large number of young children. Radioactive caesium affects the entire body, especially the liver and spleen, while strontium can cause leukaemia.

The Chernobyl Forum, a group of UN agencies, assesses the health effects and environmental consequences of the accident and they concluded that 28 emergency workers died from acute radiation syndrome and 15 patients died from thyroid cancer. They estimate that, all told, cancer deaths caused by the accident might reach 4,000 (most being thyroid tumours, for which the survival rate is almost 99 per cent) among the 600,000 clean-up workers or 'liquidators' who received the greatest exposures. The Forum also reports no increase in inherited birth defects or in other types of solid cancers arising from the fallout, a finding consistent with the Japanese atomic bomb data.

Perhaps the most encouraging thing to emerge from this disaster is the recovery of the natural world inside the exclusion zone. It was widely assumed that this area would become a desert for centuries, but in fact it now hosts great diversity of both animals and plants and all studies seem to agree that there has been no significant long-term effect, despite the fact that general levels of radiation are far higher than the normal. There are detectable effects – for example, some insects have a shorter lifespan and are more susceptible to parasites, and some birds have higher levels of albinism and, in highly contaminated areas, physiological and genetic alterations. Frogs in the zone are now darker, perhaps representing the evolution of extra protection against radiation.

More typical, however, is the comparison of vole populations with those in uncontaminated areas, for which a lifetime exposure to 10 mSv of radiation a day caused no detectable DNA damage. All the populations studied maintain stable and viable populations inside the exclusion zone – a somewhat surprising outcome but one suggesting that exposure to low levels of radiation may activate DNA repair mechanisms that confer protection.

Fukushima

There have been over 100 recorded nuclear power plant accidents in the last 70 years, the latest major disaster being at Fukushima in Japan in 2011 when a massive earthquake and associated tsunami released radioactive water into the ocean and radioactive steam into the atmosphere. Although about 50 people were reported to have been injured at the nuclear facility, the number of direct deaths from the incident is reported as zero. One thousand are said to have died during evacuation procedures, and the tsunami killed almost 16,000 people.

The vastly different consequences of the Chernobyl and Fukushima disasters appear to be due to two main factors, the superiority of the design of Fukushima's reactors and the rapid response of the Japanese government by comparison with that of the Soviet Union, which took three days to admit that an accident had occurred and did so only when radioactive sensors at a Swedish plant had been triggered by dispersing radionuclides.

Ultraviolet Radiation

Ultraviolet radiation carries sufficient energy to break the strongest (i.e., covalent) chemical bonds and thus damage biological molecules. Fortunately, almost all the radiation reaching the Earth from the Sun is absorbed by the ozone layer, but the small amount that does get through is important because sunlight drives a step in the production of vitamin D in the skin, deficiency of which causes a high proportion of premature deaths and has been linked to many diseases, including cancers.

Ultraviolet radiation is therefore both a carcinogen and a requirement for good health, and to facilitate this balance melanin pigments are made in the skin to absorb UV radiation. The cells involved are melanocytes – also present in the eye and the bowel – and they make both black eumelanin and a red pigment, phaeomelanin. We get suntanned because our skin makes eumelanin as protection against UV light. Dark-skinned people make more eumelanin and are thus better protected against skin cancer. The evolution of lighter skin as populations migrated to northern climes may have been a response to lower levels of sunshine to boost vitamin D production. Consistent with this notion is the increasing incidence of rickets (one of the most frequent childhood diseases in many developing countries as a result of malnutrition) in more recent Indo-Asian immigrants to the UK, particularly vegetarians. The predominant cause is vitamin D deficiency due to poor diet and insufficient exposure to sunlight.

Melanocytes have come to prominence in the cancer field in recent years because their uncontrolled growth gives rise to a malignant tumour called a melanoma, a relatively rare form of skin cancer – non-melanoma skin cancers are the most common – but it is the most serious. It causes three-quarters of all skin cancer deaths and each year in the UK there are over 16,000 new cases and around 2,300 melanoma deaths. The US figures are over 100,000 new cases and nearly 7,000 deaths. In both countries melanoma is one of the most common cancers in young adults. Ultraviolet exposure appears to be a major cause, but because malignant melanomas develop from moles on our skin they are the easiest cancers to detect at an early stage. Any such skin mark that changes colour, size or shape should prompt seeking medical advice. If identified early, melanomas can be treated by surgery alone with a high level of success. We will consider chemotherapy for melanoma in the next chapter.

Low-Frequency Magnetic Fields

In most countries national electrical power systems run at a frequency of 50 Hz and at about 230 volts (in North America it's 60 Hz and 120 volts). Whenever we are indoors we are almost always surrounded by a network of wires carrying alternating currents producing corresponding electric and magnetic fields. You may escape by going for a walk in the country, but bear in mind that overhead transmission lines operate at several hundred kilovolts and they, together with substations, overhead electrified railways, etc. are all radiating electromagnetic fields (EMFs).

Periodically the media have raised the question of whether these EMFs play any role in cancer, a topic made more emotive by the fact that a few reports have incriminated EMFs in childhood leukaemia. This is a very rare condition, with about 300 new cases a year in the UK and about 2,200 in the USA. There have been many epidemiological and experimental studies, but it is a very intractable problem, mainly because the magnetic fields involved are extremely weak and there is no established mechanism by which they can affect our bodies. Public concern has led to the establishment of independent bodies both in the USA (the EMF RAPID Program) and the UK (the EMF Trust) to support high-quality research into EMF effects. Thus far, the upshot of these efforts is that there is no convincing (i.e., consistently reproducible) evidence that EMFs could cause cancer.

High-Frequency Magnetic Fields: Mobile Phones

In a different range of the electromagnetic spectrum similar concerns have been raised about whether mobile (cell) phones can cause cancer, in particular two types of brain tumour (acoustic neuroma and glioma), and whether youngsters might be particularly susceptible. Mobiles work in the ultra high frequency (UHF) range (300–3,000 MHz) where the energy is a tiny fraction of what's needed to break the weakest chemical bonds, the most likely way of producing a biological effect. This means that, as with the EMF question, it's difficult to design experiments to detect effects and the cancers on which attention has focused are rare – 1 in 100,000 for acoustic neuroma and 1 in 30,000 for glioma in adults, and they are even rarer in children.

The main approach has been to ask whether mobile users get cancer more often than non-users. Even this is fraught as you need tens of thousands in two groups (users and non-users) to get statistically solid results. Unsurprisingly, the results haven't been absolutely clear-cut, but the upshot of several studies has led the UK Department of Health to pronounce that 'the current balance of evidence does not show health problems caused by using mobile phones'. They add, however, that children should be discouraged from making non-essential calls and adults should 'keep calls short'.

The exhaustive UK government-commissioned Stewart Report does, however, note one health risk from mobile phones, namely when they're in the hands of motor vehicle drivers. Interestingly, the risk seems undiminished by the use of 'hands-free' sets.

Box 8.1 Acoustic Neuromas and Gliomas

Acoustic neuromas (more accurately vestibular schwannomas) are tumours in the cells that wrap around the auditory nerve. They are benign but can nevertheless be life-threatening if they become big enough to pressurize adjacent normal tissue. Gliomas are tumours of non-neuronal cells in the brain and central nervous system.

Radon

It's a little-known fact that in most parts of the world the bulk of ionizing radiation comes from radon, a product of radium. It's an inert gas with a half-life of four days and arises from radioactive decay of uranium-238, which is present throughout the Earth's crust. The WHO estimates that exposure to radon causes tens of thousands of deaths from lung cancer each year globally, with smokers being more vulnerable. The gas normally only accumulates to significant levels inside buildings, preventable by installing sealed membranes at ground level. Although there is a statutory level for application of this measure (200 Bq per cubic metre), it's estimated that 85 per cent of deaths caused by radon arise from chronic exposure to lower levels for which preventive measures are not legally required.

Stress

As we all know, stress is a somewhat nebulous concept, but one measure is the amount of cortisol in blood or saliva. This steroid hormone is released from the adrenal gland in response to signals from the brain and its level is normally highest early in the morning, declining during the day (diurnal variation). It acts to provide energy when required by turning on the breakdown of fat and proteins, raising blood sugar levels and blood pressure. Cortisol levels are increased by food, fasting, exercise or stress, but it can also indirectly increase appetite and promote fat deposition. Seemingly two of the most stressful things are giving a speech and doing mental arithmetic in front of an audience. Five minutes of either, according to a trial at the University of Trier, is enough to push our salivary cortisol levels up two- to four-fold.

There's evidence that some breast cancer patients may have abnormal cortisol profiles (i.e., raised levels or abnormal variation) and that this associates with shorter survival (3.2 vs 4.5 years). Suppression of the immune system by raised cortisol levels – reduced numbers of white cells in the blood – may be a factor. Another strand of cortisol involvement is the evidence from several studies that night-shift work, which disrupts normal diurnal rhythms, is associated with increased incidence of breast cancer. The suppression of melatonin levels may be involved, as may perturbed cortisol variation. Notwithstanding these suggestive findings, cortisol-type steroids can suppress the growth of some tumours and have been used in chemotherapy.

Where Do We Stand and What Can We Do?

At the outset we defined three categories of cancer causes: replication errors, inherited mutations and environmental factors. Our survey of possible environmental causes has shown that they more or less fall into two groups: those for which the evidence is incontrovertible and those for which it isn't. For smoking there is a combination of overwhelming epidemiological evidence coupled to equally strong molecular data. More generally it is clear that what we eat and drink and excessive ionizing radiation (and UV radiation) are also major factors.

With that in mind, let us tackle the crunch question: 'How can we reduce our risk?' The answer turns out to be easy: apply the rather old-fashioned approach of common sense.

First, forget about all the things over which we have no control and concentrate on the others. For that we can do no better than tap the distilled wisdom of the American Institute for Cancer Research and the World Cancer Research Fund and pass on this good advice – ignoring Oscar Wilde's comment that 'It's the only thing to do with it as it is never of any use to oneself':

1. Eat a balanced diet containing plenty of vegetables, fruits, lentils, beans and whole grains such as brown rice and whole wheat pasta (at least two-thirds of any meal). These have a relatively low calorie content but their water and fibre content fill you up and, of course, they give you most of the vitamins and minerals you need.
2. Stay as lean as possible (without becoming underweight). A minimum of 30 minutes of physical exercise a day will also help to prevent excessive body fat from forming, an established risk factor for cancers of the oesophagus, endometrium, pancreas, kidney, bowel and breast in post-menopausal women. For prostate cancer, too, there is evidence that physical occupations carry a substantially lower risk than desk jobs.
3. Don't eat too much red meat (no more than 500 grams cooked weight a week). Some red meat is good because it contains nutrients, but other components can damage the lining of the bowel.

4. Don't drink alcohol or anything with a high sugar content. If you can't face an alcohol ban (and it is regularly reported to protect us against heart disease), limit yourself to one (women) or two (men) drinks per day.
5. Don't eat too much salt.
6. Don't use food supplements unless specifically suggested by your physician, and check that they aren't profiting from the sales.

On the global front, the three most effective anti-cancer things we could do are (1) stop use of tobacco; (2) provide clean water for everyone; and (3) reduce consumption of red meat. Collectively that would more than halve cancer incidence.

9 Treating Cancer by Chemotherapy

Paul Ehrlich's pursuit of drugs to combat infectious diseases led him to the notion of a 'magic bullet' – something that could kill microbes such as bacteria without any harm befalling the infected individual. He also came up with the word 'chemotherapy' to mean the use of chemicals to treat disease. Having chronicled the rise in the profile of microbiota in cancer, it might be informative, at the outset of this chapter, to refer to a striking demonstration of the challenge presented by bacteria because it has a strong parallel with cancer.

The golden age launched by Alexander Fleming's celebrated discovery of penicillin in 1928 is long gone and the last antibiotic class to become a successful treatment was discovered in 1987. While the discovery curve has drifted ever downwards since 1960, the bugs have been busy. Just how busy a bug can be was shown by a large-scale experiment carried out by Roy Kishony and friends at Harvard Medical School. They built a 'Mega-Plate' – a Petri dish 2 feet by 4 feet filled with a jelly for the bacteria to grow in. The bugs were seeded into channels at either end so they would grow towards the middle. The only thing stopping them was four channels dosed with anti-biotic at increasing concentrations – 10 times more in each successive channel.

The bugs grew until they hit the first wall of antibiotic. This stopped them in their tracks, but after a while clusters of cells began to move into the first drug channel. Gradually other groups followed until a tidal wave swept over that barrier. This was repeated at each new 'wall' – four times until the whole tray was a bug fest. When they paused at each new 'wall' the bacteria were picking up random mutations in their DNA until they were able to deal with the higher drug environment. This experiment was a fantastic, visual display of the rise of drug resistance to once-effective agents. And it's terrifying because it took about 11 days for them to overcome four levels of drug. It's even more scary in the speeded-up movie as that lasts less than two minutes. Bacteria that acquire resistance to several antimicrobials – 'superbugs' – are 'multidrug resistant'.

Sound Familiar?

It should do. Each advancing front is a colony that has acquired mutations enabling it to advance. But you could perfectly well think of these as clones of cancer cells acquiring mutations in key 'driver' genes that advance the evolution of the tumour. That's pretty scary too, and the only good news is that animal cells reproduce much more slowly than bacteria.

So the fields of antibiotics and cancer therapy face a similar challenge – to come up with new and effective drugs to control unwanted cell growth. However, before we look at the current cancer options we need a word about detection. The earlier we can reliably determine the presence of a cancer, the greater the chances of successful treatment, but detection raises the question of the sensitivity and specificity of tests. Sensitivity is the capacity of a test to identify accurately all patients with a condition. Specificity is the ability of a test to identify accurately people without the condition. The perfect test is one that has 100 per cent sensitivity and 100 per cent specificity (i.e., in a random population the test accurately diagnoses all patients that have a condition and also identifies accurately all patients without that condition). In cancer diagnosis there are no perfect tests – everything has some margin of error and a balance must be struck between sensitivity and specificity. As a general rule, more invasive tests have improved sensitivity and specificity, but are less acceptable to patients and are only performed if essential.

Screening

In 1928 the Greek pathologist George Papanicolaou and the Romanian physician Aurel Babes reported that cancer cells could be detected in vaginal smears after staining with combinations of dyes. What has become known as the Pap test came into widespread use in the 1950s and continues to be recommended in many countries for women from about 20–50 years of age. It remains the model for cancer screening and through the early diagnosis of cervical cancer has saved the lives of millions of women.

Setting aside Pap staining and visible skin cancers, cancer screening currently depends on detection of tumour biomarkers or on medical imaging. Biomarkers are molecules or cells detectable in the circulation. Perhaps the most familiar is prostate specific antigen (PSA), raised blood levels of which would prompt an MRI (magnetic resonance imaging) scan before testing the tissue directly by a biopsy. However, 7 in 10 men with raised PSA levels will not have cancer, and more than 1 in 5 of those with prostate cancer will have normal circulating levels of PSA. Cancer antigen 125 (CA125) is considered a biomarker for ovarian cancer, but only about 50 per cent of these cancers release increased amounts of CA125 at an early stage.

These examples illustrate the current cancer predicament over biomarkers. The PSA test has FDA approval, but high false positive and false negative rates limit diagnostic usefulness. CA125 is similarly limited, although both PSA and CA125 can be used to monitor treatment responses. The current limitations of biomarker detection mean that there are no national screening national programmes for prostate or ovarian cancer in the UK, although a trial is underway of MRI scanning for prostate cancer.

In the UK all people aged 55 are now invited for a one-off colonoscopy and a home testing kit for faecal blood is sent every two years to those between 60 and 74 years of age to screen for bowel cancer. Among various trials, University College London is running the SUMMIT study of early lung cancer detection using low-dose CT (computerized tomography) and trialling a new blood test that aims to detect circulating tumour DNA in the blood for the early detection of multiple cancer types.

Mammography

Notwithstanding the improvement in breast cancer survival in many countries over the past few decades, early detection remains a pressing and controversial issue. Screening for cancer is a seductive notion because intuitively it would seem that, however imperfect the method, it can only do good. However, for breast cancer screened by low-dose X-rays or, more recently, by digital X-ray tomosynthesis, which is similar to CT but gives a series of slices through the breast rather than a 3D image, analysis of the results reveals that mammography probably does more harm than good.

Rather than consider in detail even a few of the huge number of mammography studies, it is more informative to go straight to the Cochrane Collection's distillation of trials that have involved more than 600,000 women. Their synopsis emphasized the variation in quality between different studies and made the point that the most reliable showed that screening *did not* reduce breast cancer mortality. One example of how a trial might be less reliable arises from the problem of deciding *without bias* whether a woman is assigned to the group that undergoes mammography or to the other, unscreened, group. The key facts emerging from this review were that for every 2,000 women invited for screening over 10 years, one will avoid dying of breast cancer and 10 will be treated unnecessarily. In addition, false alarms will subject 200 women to prolonged distress and anxiety.

These conclusions were, as one would expect, consistent with several single studies seemingly conducted to a high standard. Thus, for example, the Swiss Medical Board in 2014 recommended that the current screening programmes in Switzerland should be phased out and not replaced. A Danish study concluded that decreases in deaths from breast cancer were more likely to be the result of changes in risk factors and improved treatment, rather than screening, and the *British Medical Journal* commissioned an independent look at UK data that also concluded that the benefit of screening was very small.

The idea of abandoning X-ray screening for breast cancer was met with predictable global outrage and mammography remains available in Switzerland. In England, the recommendation is that women aged 50–70 are offered mammography every three years. The US Preventative Services Task Force 2016 suggests mammography every two years between the ages of 50 and 74, concluding that 'the benefit of screening mammography outweighs the harms by at least a moderate amount from age 50 to 74 years and is greatest for women in their 60s'. All of which suggests that there is much scope for improving cancer detection. Thus, for breast cancer, patients with a family health history are tested for mutations in major drivers (e.g., *BRCA1*, *BRCA2*, *TP53*) and those carrying *TP53* mutations can be offered MRI screening from age 20. Corresponding flexible approaches should be developed for biomarkers such as cell-free DNA (Chapter 10).

Diagnosis, Staging, Grading and Monitoring

The next question is how is a cancer diagnosed? A screening test usually leads to diagnostic testing and staging. In the majority of cases definitive diagnosis comes from direct tissue sampling (usually in the form of a biopsy or surgical resection) and subsequent histopathological analysis – what the cells look like under a microscope after a variety of different stains have been applied – while staging is performed using medical imaging. For most tumours a histopathological grade is given to the tumours (in general a high grade means cells are very abnormal and hence a bad prognosis; a low grade means cells are only slightly abnormal, hence a better prognosis).

Imaging

Today virtually every cancer patient will have imaging performed regularly using one or a combination of CT, MRI, ultrasound and positron emission tomography (PET). The development of these techniques is due to wonderful advances in chemistry, physics and computing, and they have played a huge role in improving survival for most cancers. The underlying physics is complex and beyond the remit of this book, and indeed beyond many in the field – the next time you meet a radiologist (image analysis expert), ask them for a clear explanation of MRI or CT image reconstruction – and prepare to be confused! Suffice to say that CT, MRI, PET and ultrasound are all capable of producing 3D images of the internal structures of the body. To do this, CT uses X-rays, MRI uses the magnetic properties of some atoms, usually protons (H^+, of which there are two in every water molecule so most clinical MRI scans are essentially maps of water concentration in the body), ultrasound uses the reflection of high-frequency sound waves and PET detects the location of radioactive isotopes injected into the body.

Medical imaging began in 1895 when Wilhelm Röntgen accidentally imaged his hand using X-rays (Figure 9.1). In the last 50 years the aforementioned imaging techniques have been transformed from grainy, barely recognizable images to 3D maps of the body with sub-millimetre resolution.

The capacity to stage cancers by imaging – that is, to determine whether they have spread locally, metastasized, continued growing or responded to treatment – has revolutionized oncology and most treatment decisions are guided by imaging. Broadly speaking, localized tumours are treated with surgery or radiotherapy while tumours that have spread are treated with chemotherapy.

There are two main types of radiation therapy and both use ionizing radiation (high-energy particles or waves) to kill cancer cells. Recall that in Chapter 8 we noted that ionizing radiation has sufficient energy to release electrons

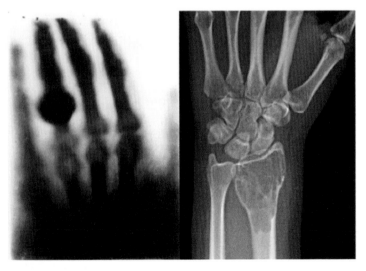

Figure 9.1   Over 125 years of imaging. Left: the first medical X-ray by Wilhelm Röntgen of his wife Anna Bertha Ludwig's hand with rings. Right: a modern wrist X-ray showing a large giant cell tumour (benign) in the distal radius.

from atoms (i.e., to ionize them). Any biological molecule can be damaged by ionizing radiation, but the most significant target is DNA, the aim of radiation therapy being to inflict sufficient damage to the genome that the cell cycle cannot continue and cell division is blocked.

More than half of cancer patients receive some form of radiation therapy and it is used to treat most types of cancer and some non-cancerous (benign) tumours. Radiotherapy is also used before surgery to shrink a tumour (neoadjuvant therapy) or after surgery to prevent re-growth of residual cancer cells (adjuvant therapy). In combination with chemotherapy, radiotherapy can alleviate the symptoms of advanced cancers. The level of radiation required typically ranges from 20 to 80 Gy. This is about 100,000 times the dose from an abdominal X-ray and 10,000 times that of a typical CT scan.

Brachytherapy is a form of radiation therapy in which a sealed radiation source is implanted next to the target and is often used to treat cancers of the head and neck, breast, cervix, prostate, oesophagus and eye, the commonly used isotope being iridium-192 (^{192}Ir). It has the advantage that by comparison with external beam radiotherapy it minimizes radiation exposure of healthy tissue.

Radiotherapy is reliant on the imaging delineating tumours accurately so that treatment can be designed to maximize the dose to the tumour and minimize that to surrounding normal tissues. However, knowing the location of a tumour is only part of the challenge and, as demonstrated in the preceding chapters, knowing something about the underlying biology – which mutations are present and which pathways are working overtime – is important for designing targeted chemotherapy.

Currently biopsy is indispensable for this purpose, but it is an invasive procedure with a small but not negligible risk of complications – approximately 1 in 1,000 patients die as a result of a lung biopsy and it is therefore not something to be frequently repeated. Biopsies only sample a very small proportion of the tumour, which gives a selective picture because tumours and their metastases are not homogeneous, unchanging balls of cells but are dynamic organs often rapidly evolving and containing multiple different types of cells. In addition to providing high-resolution anatomical images to delineate the position of a tumour and its size, imaging can also provide information about the underlying tumour biology. PET achieves this by detecting the location of injected radionuclides within the body. The most commonly used is fluorine-18 deoxyglucose (FDG), a glucose molecule with an oxygen atom replaced by a fluorine-18 radioisotope. It's taken up by cells, phosphorylated and trapped. The scanned images that can be generated are essentially maps of glucose uptake, and because most cancer cells take up lots of glucose they tend to light up as hot spots. Currently in clinical practice, FDG-PET is almost entirely used for the purpose of increasing the sensitivity of CT and MRI for detecting and differentiating tumours from normal tissue. But the uptake of FDG itself conveys information about the biology of the tumour: if FDG is taken up the tumour has a good blood supply, an important consideration for drug treatment.

Numerous biomarkers based on functional imaging are at various stages of development, and these are not merely limited to radiotracers and PET imaging. MRI can now follow some metabolic reactions in the body – tracking the conversion of an injected molecule into metabolic products as the result of enzyme reactions. This offers the advantage of providing information about what the cells are actually doing metabolically to the point of being able to estimate the rate of metabolic reactions, rather than merely measuring the static accumulation of metabolites. It is anticipated that this type of imaging data will improve current diagnosis, staging, grading and treatment response detection. The current sensitivity of imaging for solid tumours is high and most cancers are first identified using these methods. However, imaging is by no means perfect, and tumours are frequently missed, particularly in the early stages. It may seem difficult to miss an abnormal lump in the body, but recall that all tumours start as normal tissue so there is a sliding scale from what is definitely normal to what is definitely tumour. It seems likely that the importance of imaging as a screening tool will decline as other biomarkers, particularly those derived from blood tests, are developed. These will be discussed in the next chapter.

Chemotherapy for Cancer

In Chapter 2 we charted the advance of surgery as the main treatment for cancer well into the twentieth century, and it remains effective for many types of cancer. It was joined by radiotherapy from the 1890s, when X-rays were first used to treat cancer. Both of these fields have seen startling technical advances in recent years, exemplified by a method called SpiderMass that uses a laser microprobe to sample tissue during tumour surgery. The probe is directly linked to a mass spectrometer so that the tissues are subjected to molecular analysis in real time during the operation, the data guiding the surgeon in distinguishing tumour from normal tissue. In radiotherapy the last decade has seen the CyberKnife System come into use as the first fully robotic radiotherapy device. It uses X-ray cameras to monitor the position of the tumour and stereotactic body radiation to direct precise doses with extreme accuracy, its robot control adjusting for patient movement in real time.

Despite these phenomenal innovations, it has long been clear that controlling cancers, particularly advanced cases with metastases, cannot be achieved by surgery alone, no matter how sophisticated. It is going to depend on drugs – chemotherapy. The term chemotherapy can cause confusion, but it simply means the use of drugs to destroy cancer cells. It therefore includes small molecules, both naturally occurring and manmade, as well as monoclonal antibodies and therapeutic viruses, both of which are engineered. Chemotherapy is frequently used in combination with surgery or radiotherapy, when it's known as adjuvant chemotherapy. Neoadjuvant chemotherapy means drug treatment to reduce the size of primary tumours *before* surgery or radiation treatment. Palliative chemotherapy is the use of drugs to alleviate symptoms and perhaps give some extension of life. The drugs are normally administered systemically, so that they circulate in the bloodstream and can affect every cell in the body.

A New Era

We have already mentioned the first attempts at chemotherapy – the use of arsenic by the Egyptians and herbal remedies in China – and we traced that history to the 1940s and to Charles Huggins. His demonstration that the growth of prostate tumours could be controlled by hormonal treatment could be said to have launched the modern era of cancer drug therapy. In the same decade, Gustaf Lindskog, an American thoracic surgeon, together with Louis Goodman and Alfred Gilman, carried out the first pharmaceutical (i.e., manufactured chemical) treatment for cancer in which administration of nitrogen mustard produced marked regression of lymphomas. The nitrogen mustards are based on a small chemical modification of mustard gas, one of a number of poisons synthesized by the German chemist Fritz Haber, best known for devising a method for making ammonia-based fertilizer. Victims of mustard gas, used as a weapon in the First World War and also released from a bombed ship in Bari harbour, Italy in the Second World War, had severely depleted white blood cell counts and loss of bone marrow and lymphatic tissues. Trials of nitrogen mustards as anti-cancer agents in animals and humans began in 1942, and these led to the development of agents that add an alkyl group to DNA, thereby preventing cell replication. Of these, cyclophosphamide has become the most widely used.

Inhibiting Proliferation

These early efforts led to the synthesis of several related alkylating compounds that remain in use today. These include chlorambucil, used to treat chronic lymphocytic leukaemia and lymphomas, and cyclophosphamide, commonly used in combination with other agents against a range of cancers. Starting with penicillin, antibiotics were also screened for anti-cancer effects, leading to actinomycin D and other anti-tumour antibiotics still in use. This family of agents works in various way to block DNA replication and RNA synthesis and hence cell division.

The next major contribution came from the Buffalo-born pathologist Sidney Farber, thought of as the father of modern chemotherapy. In 1948 he discovered that folic acid is critical for the proliferation of cancer cells in leukaemias. Folic acid (folate or vitamin B9) is an essential cofactor for enzymes that synthesize the bases of DNA and RNA. The Indian biochemist Yellapragada Subbarow synthesized aminopterin, a derivative of folic acid, that competes with folate in binding to a key enzyme (dihydrofolate reductase), thus depleting the pool of bases for DNA and RNA. Farber showed that aminopterin could give temporary remission of acute lymphoblastic leukaemia in children – the first drug with activity against liquid tumours. Aminopterin was supplanted in the 1950s by more effective agents – methotrexate, 6-thioguanine and 6-mercaptopurine, all of which block DNA synthesis. They are not specific, targeting any dividing cell, which is why they also suppress the immune system. Despite these limitations, in 1956 methotrexate became the first successful drug treatment for a metastatic cancer when it was used against a rare tumour called choriocarcinoma that spreads to the lungs.

This decade also saw the development of 5-fluorouracil (5-FU), a compound that also blocks DNA replication (it inhibits the enzyme thymidylate synthase, blocking synthesis of thymidine). 5-FU came into clinical use in 1962. It is active against a range of solid tumours and remains the cornerstone of colorectal cancer treatment. The 1960s also saw the introduction of combinations (cocktails) of drugs, showing to spectacular effect that several agents acting simultaneously could be much more effective than any one alone. Thus, for childhood acute lymphoblastic leukaemia (ALL) the combination of vincristine, methotrexate, 6-mercaptopurine and prednisone lifted the remission rate from essentially zero to 60 per cent, and today about 90 per cent of children with ALL can be cured. Combination chemotherapy has had similar dramatic effects on childhood acute myeloid leukaemia (AML), now with survival rates of 65–70 per cent, and on acute promyelocytic leukaemia (APL), a sub-type of AML that is essentially curable. Combination strategies have made similar impacts on Hodgkin's disease and testicular cancer.

The idea is that targeting the process of cell division at several points, together with suppressing the immune response and inflammation (e.g., with prednisolone), overcomes the adaptability of the tumour cell.

An obvious alternative to blocking DNA replication as an anti-cancer strategy is to inhibit the cyclin-dependent kinases (CDKs) that, as we saw in Chapter 5, are sequentially activated to drive the cell division cycle. Much effort has gone into this approach, and in 2015 the US FDA approved the first drug of this type, palbociclib, a kinase inhibitor, for use in postmenopausal women with oestrogen receptor-positive and HER-negative breast cancer. The inhibitory effect of palbociclib blocks RB1 phosphorylation, preventing the cell from passing the checkpoint at the exit of G1 and progressing through the cell cycle. At least 12 CDK inhibitors have entered trials, but as yet they have not been notably successful, possibly because, as suggested by gene knock-out experiments in mice, loss of a specific CDK does not affect cell proliferation – perhaps another example of the adaptability of cells.

Selective Oestrogen Receptor Modulators

We've already met the steroid hormones – corticosteroids and sex steroids – that have multiple effects on cell behaviour. They're important in cancer because oestrogens stimulate the growth of breast and endometrial tumours while androgens (including testosterone) stimulate prostate cancer cells to grow. The levels of steroid hormone receptors are used to predict response to endocrine therapy (also called hormone therapy – that is, treatment that modulates the effect of hormones). About two-thirds of breast cancers are ER (oestrogen receptor) and/or PR (the female sex hormone progesterone) positive. ER^+/PR^+ tumours respond much better to hormone therapy than do ER^+/PR^- tumours.

The role of oestrogen in a variety of diseases, including cancers and osteoporosis, has prompted the development of selective oestrogen receptor modulators (SERMs), a class of drugs that act on the ER. SERMs attach to the hormone receptors and either mimic the action of the natural hormone or prevent it from acting. There are three SERMs: raloxifene, toremifene and tamoxifen. Raloxifene acts like oestrogen in bone but blocks oestrogen action in breast tissue and in the uterus. Thus, it's used to treat osteoporosis in postmenopausal women and to reduce the risk of breast cancer. Toremifene is oestrogenic in bone and the uterus but anti-oestrogenic in breast, and is used to treat advanced breast cancer in postmenopausal women.

First made in 1962 in a search for a morning-after contraceptive pill, tamoxifen has been the standard anti-oestrogen therapy for breast cancer since 1980 because it can antagonize the action of oestrogen. It's a 'pro-drug', meaning that it is taken by mouth and then metabolized by the body to active forms that compete with oestrogen binding to ERs. It is 'selective' because its action depends on the tissue and it is a first-line hormonal treatment of ER-positive metastatic breast cancer. In breast cells ER/tamoxifen complexes block transcription of genes normally switched on by oestrogen (acting as an agonist). In the endometrium it acts as a less effective version of oestrogen (it's a 'partial agonist'), but in bone it is an ER agonist – that is, it mimics the effects of oestrogen and prevents bone loss.

Oestrogens are synthesized from androgens by the enzyme aromatase and in the last 15 years aromatase inhibitors (anastrozole, letrozole and exemestane) have been introduced as anti-cancer agents that work by lowering the level of oestrogen. Fulvestrant is an oestrogen receptor antagonist – it blocks the action of oestrogen on cells and causes the cell to degrade and digest the oestrogen receptors – it's a selective oestrogen receptor degrader (SERD).

The above survey highlighted some major milestones in the field of anti-cancer drugs since the 1940s, but it is notable that, although the agents are a mixed bunch, they have in common the target of inhibiting DNA synthesis, cell cycle progression or transcription of key genes. In other words, they take aim at the heart of cellular life by stopping cell proliferation – the first of the classes of targets for anti-cancer drugs represented in Figure 9.2. This scheme is based on the *Hallmarks of Cancer*, defined in an influential review by two leading figures in cancer research, Doug Hanahan and Bob Weinberg, that set out the features distinguishing a cancer cell from its normal counterpart. Figure 9.2 shows the nine broad categories of molecular events that we've described as contributing to cancer development, and we will use it as the

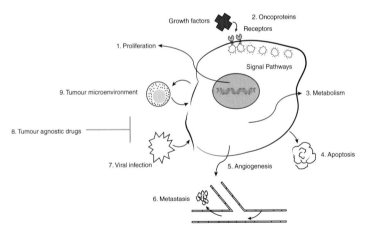

Figure 9.2 Classes of targets for anti-cancer drugs. (1) Proliferation, (2) oncoproteins (growth factors, receptors, signal pathways); (3) metabolism; (4) apoptosis; (5) angiogenesis; (6) metastasis; (7) infection by oncogenic viruses; (8) tumour agnostic drugs; and (9) the tumour microenvironment.

Box 9.1 Drug Nomenclature

Names that end in -mib and -nib:

  -tinib: tyrosine kinase inhibitors

  -zomib: proteasome inhibitors

  -ciclib: cyclin-dependent kinase inhibitors

  -rafenib: BRAF kinase inhibitors

  -parib: PARP inhibitors

basis for selecting specific examples of chemotherapy from the over 400 different anti-cancer drugs and drug combinations currently in clinical use.

Oncoproteins: Growth Factors, Receptors, Signal Pathways

Hard on the heels of the full sequencing of the human genome in 2003, Michael Stratton, working at the Wellcome Sanger Institute in Cambridge, completed a study that utilized both the fantastic technical developments of the project and the database of the sequence itself. Five hundred genes were selected, each of which encoded a member of the kinase enzyme family thought to be involved in signalling pathways that controlled growth. A large number of tumour samples were sequenced and analysed for mutations in the kinases. Unsurprisingly, a number of kinases already known to be mutated in human cancers were detected. What was a surprise, however, was finding the kinase gene *BRAF* to be mutated in about two-thirds of the melanomas. The mutation was of a single base that changed just one amino acid. This 'point mutation' switches on its activity just as activating mutations do for RAS proteins (Chapter 6). This was an important finding because melanoma, although not one of the most prominent cancers, killed over 60,000 people in 2018 and until the Sanger group's discovery virtually nothing was known about it at the molecular level. They had discovered a new 'cancer gene' that played a major role in a very prevalent cancer.

Almost more remarkable, within a few years a drug had been produced that is very efficient at blocking the action of the mutant form of BRAF (i.e., it is tumour-cell specific) and thus offers a chemotherapeutic approach to treating melanoma. Vemurafenib (named from 'V600E mutated BRAF inhibition') received FDA approval for the treatment of late-stage melanoma in 2011.

Vemurafenib has turned out to be highly instructive in the matter of designing specific agents to combat cancer. The initial results showed striking remissions of melanomas that were otherwise untreatable, but it gradually emerged that most responses are transient and, although some patients do have long-lasting responses, the majority relapse within one year. This failure is not because the drug has stopped working: it's due to the appearance of drug-resistance mechanisms that short-circuit the drug target and restore MAPK pathway activity. More generally, the variety of these adaptations utilized by cells to neutralize the effects of drugs is truly astonishing and they are a major limitation on the effectiveness of chemotherapy. Thus, for example, methotrexate can be rendered ineffective by amplification of the gene encoding its target protein, so that eventually the maximum dose that can be administered is reached. Tamoxifen can be neutralized by loss of the oestrogen receptor. In addition, drug efflux, mediated by trans-membrane pumps, may be enhanced to export the drug from cells. Drug resistance is very variable, with some tumours having generally low levels of resistance (e.g., Hodgkin's lymphoma, childhood acute leukaemia) while others usually respond to initial treatment but eventually acquire resistance (e.g., non-small-cell lung cancer) and some types are inherently resistant (e.g., melanomas that have the highest mutation load of any cancer). In summary, it's an example of the blancmange effect that we described in Chapter 5. For vemurafenib these adaptations have prompted a number of trials of combinations with other MAPK pathway inhibitors and with immunotherapy, and we'll return to these in the next chapter.

Kinase Inhibitors

Since the SRC protein was revealed as a kinase that could be activated by mutation, it's become clear that phosphorylation is a central theme of cell control and, predictably, numerous kinases have now been identified as 'oncoproteins' – that is, they can acquire mutations that permanently switch them on as cancer drivers. The human genome encodes almost 100 tyrosine kinases, of which over 50 are receptor tyrosine kinases. We met one, EGFR, when talking about mutations in Chapter 7.

Much effort has gone into drugs that block kinases and some 50 small molecule protein kinase inhibitors have received FDA approval. The first such inhibitor (approved in 2001) targets the mutant BCR–ABL1 protein that is expressed only in tumour cells. Imatinib is the first-line treatment for chronic myelogenous leukaemia (CML) and it has had the striking effect of more than doubling the five-year survival rate from 31 per cent in the early 1990s to 69 per cent since 2009. However, like vemurafenib, the story of imatinib is a cautionary tale in that, after the initial response, small clones of cells emerge with a mutated version of BCR–ABL1 that does not bind imatinib and these eventually become dominant and confer resistance. This led to a second-generation inhibitor, dasatinib (approved in 2006), that counters this problem effectively, perhaps because it is less specific than imatinib – that is, it blocks a number of other kinases as well as BCR–ABL1.

The EGFR is overexpressed or otherwise mutationally activated in a wide variety of tumours, notably in glioblastoma and lung and breast cancers, with one result of sustained EGFR activity being abnormal stimulation of a signalling cascade via RAS that attenuates cell death (apoptosis). Two of the most successful targeted cancer therapies to date, used for treating non-small-cell lung cancer and other carcinomas, are gefitinib and erlotinib, small molecules that inhibit the kinase activity of mutant forms of EGFR. In addition to small molecules, antibodies have also been developed that target the EGFR, notably cetuximab, which is used for metastatic bowel cancer. Herceptin targets HER2, a close relative of EGFR, and can be used alone or with other agents in the treatment of breast cancer and stomach cancer. The first drug based on a monoclonal antibody, rituximab, come into use in 1997. It binds to a protein (CD20) found mainly on the surface of B cells and is used to destroy overactive B cells or dysfunctional B cells in leukaemia and lymphoma cells.

Box 9.2 Antibodies

An antibody (or immunoglobulin) is a large, Y-shaped protein made by white blood cells (B cells) in the immune response that neutralizes pathogens – bacteria and viruses. Antibodies work by recognizing a unique molecule or molecular fragment – an antigen – from the pathogen. Monoclonal antibodies are proteins made in the laboratory by identical immune cells that are all clones of a unique parent cell, hence they all bind to the same region of an antigen (epitope). Humanized antibodies are made in the laboratory by combining a human antibody with a small part of a mouse or rat monoclonal antibody. The latter part provides the site that binds to the target antigen. The human part makes it less likely to be destroyed by the immune system.

Metabolism

When in 1924 Otto Warburg reported his finding that cancer cells consume tremendous amounts of glucose, he established a new branch of cancer therapy – although it was not to open for business for a further 70 years. It's a striking phenomenon, now referred to as aerobic glycolysis or the Warburg effect, in which tumour cells convert most of their glucose into lactate (by glycolysis), even in the presence of oxygen – rather than making energy more efficiently by using their mitochondria to make ATP. This unexpected finding provides a therapeutic opening: might it be possible to use a modified form of glucose that is taken up but cannot be further metabolized – so it stops tumour cells in their tracks? 2-deoxyglucose does indeed block glycolysis but, although it has anti-proliferative effects, neither it nor other agents targeting glucose uptake and use have been successful as treatments, possibly because the proliferation rates of many normal cells (bone marrow, intestinal crypts (folds) and hair follicles) often exceed those of cancer cells.

The perturbed metabolism of cancer cells has the effect of increasing levels of reactive metabolites, produced by metabolic enzymes. These include reactive oxygen species made by mitochondria that can react with proteins and DNA to drive oncogenesis. A number of mitochondrial inhibitors are under development, including metformin, a standard clinical drug used to treat type 2 diabetes mellitus.

There is one example, thus far unique, of somatic mutations converting a metabolic enzyme into a form that helps to drive tumour development. The enzyme is isocitrate dehydrogenase (IDH), mutated forms of which occur in a wide variety of cancers, including brain tumours (glioblastoma multiforme) and AML. These mutations cause a different reaction product to be made that can produce epigenetic changes in DNA (described in the next chapter), affecting gene expression and ultimately leading to cancer. A number of agents inhibiting the enzyme activity of mutant IDHs are in development, and enasidenib (formerly AG-221) received FDA approval in 2017 as a first-in-class inhibitor of mutated IDH2 for treating AML. A handful of other metabolism-modulating drugs have reached clinical trials.

Apoptosis

We mentioned earlier that a controlled programme of cell death is part of normal development in animals and, perhaps less surprisingly, that apoptosis is one of the best protections against cancer that we have. *TP53* is a critical regulator of apoptosis, being activated by a range of stress signals, particularly damage to DNA, and in its role as 'guardian of the genome' it can induce cell growth arrest or apoptosis. Apoptosis is carried out by a group of enzymes called caspases that cleave target proteins – essentially they inactivate hundreds of different proteins by chopping them into pieces. Considerable efforts have been made to produce drugs that either stimulate pro-apoptotic or inhibit anti-apoptotic proteins. These have included caspase inhibitors and a variety of compounds derived from plants, including quercetin, curcumin and green tea. Although this area clearly has potential, as yet it has not produced an effective, clinical anti-cancer agent.

A completely different approach has used modified viruses to destroy tumour cells that have lost *TP53*, while having no significant effect on normal cells. These include the genetically engineered adenoviruses ONYX-015 and the closely similar H101. The latter received regulatory approval in 2005 in China for the treatment of head and neck cancer. The first oncolytic virus to be approved by the FDA (2015) was talimogene laherparepvec (T-VEC made by Amgen) and it has shown some promise in treating head and neck cancer and melanoma.

A radical approach to dealing with loss of *TP53* in cancers would be to insert a normal version of the gene in defective cells. At the moment this is not feasible in humans, although we'll discuss the potential of gene therapy in the next chapter. In transgenic mice, however, it has been shown that knocking out one copy of *TP53* cuts the lifespan of the mice by about half as a result of early-onset cancers. When three copies of the *TP53* gene (rather than the normal two) are expressed, mice are significantly protected from cancer. However, adding an extra copy of *TP53* can cause premature ageing as well as conferring tumour protection, indicating that how gene expression is controlled critically affects the upshot of gene manipulation, and at the moment we cannot regulate this with sufficient precision to embark on human experimentation.

Angiogenesis

It was Judah Folkman at Harvard Medical School who first observed in the 1970s that the growth of new blood vessels from pre-existing vessels – called angiogenesis – plays a significant role in the development of many cancers. When tumours turn on angiogenesis it's sometimes called an 'angiogenic switch' and it's driven by a shift in balance between pro- and anti-angiogenic factors.

Blood vessels are lined with a single layer of endothelial cells – host cells that are genetically stable relative to cancer cells, so they should be less adept at mounting resistance to drugs. Oxygen deprivation (hypoxia) is a powerful growth stimulus for endothelial cells (and hence for angiogenesis), in part by turning on production of the signal protein vascular endothelial growth factor (VEGF). Since the 1990s a number of anti-angiogenic drugs have been produced and in 2004 the humanized version of a monoclonal antibody to VEGF, Avastin, became the first FDA-approved anti-angiogenic drug. Avastin approval was initially for metastatic bowel cancer (in combination with 5-FU) and has subsequently been extended to use against a range of other cancers. Despite its promise, Avastin received a set-back in 2010 when the FDA rescinded its approval for breast cancer because evidence had by then accumulated showing that it did not prolong life and has serious side-effects. Avastin retains approval for treating other cancers.

In the 1990s Robert D'Amato, a colleague of Folkman, showed that the sedative drug thalidomide inhibited angiogenesis and this led to the demonstration that about one-third of patients with multiple myeloma responded to the drug. Thalidomide was initially marketed as a mild sleeping pill, safe even for pregnant women, but it caused thousands of babies worldwide to be born with deformities and was withdrawn in 1961. However, in 2006 it was approved for the treatment of newly diagnosed multiple myeloma in combination with dexamethasone. It is also an effective therapy for leprosy. Thalidomide appears to act both as an anti-angiogenic and also as a positive modulator of the immune system.

Currently 10 anti-angiogenic agents have received FDA approval but their effects have been relatively limited and there is evidence that after anti-angiogenic treatment some cancers may recur in more aggressive forms. A number of other drugs have been used that target endothelial cells, including some taxanes, originally derived from plants (e.g., paclitaxel and doce-taxel). Overall, however, despite the attraction of anti-angiogenic therapy and the application of much effort, this approach to cancer has thus far been of very limited success.

Metastasis

Over a century ago there lived in London an astute physician by the name of Stephen Paget who specialized in treating breast cancer, and his observations had led him to the question of how the movement of tumour cells around the body is directed or, as he elegantly phrased it in his paper of 1889: 'What is it that decides what organs shall suffer in a case of disseminated cancer?' Paget knew that it was not simply a matter of the cells sticking to the first tissue they met because breast cancer cells often migrated to the lungs, kidneys, spleen and bone. A few years earlier the Austrian ophthalmologist Ernst Fuchs had commented that certain organs appeared predisposed to receive wandering tumour cells and this led Paget to a botanical analogy for tumour metastasis: 'When a plant goes to seed, its seeds are carried in all directions; but they can only live and grow if they fall on congenial soil.' From this emerged the 'seed and soil' theory of metastasis, its great strength being the image of interplay between tumour cells and normal cells, their actions collectively determining the outcome. Rather charmingly, Paget concluded his paper with: 'The best work in the pathology of cancer is now done by those who are studying the nature of the seed. They are like scientific botanists; and he who turns over the records of cases of cancer is only a ploughman, but his observation of the properties of the soil may also be useful.'

It's a sobering fact that Paget's aphorism of 'seed and soil' pretty well summed up our knowledge of this most important of questions until the twenty-first century. One of the reasons for the hiatus is the dependence of scientific progress on available technology – in this case genetically modified mice and antibodies tagged with fluorescent labels to detect specific proteins in cells.

Many of the key revelations over the last 15 years have come from David Lyden and his colleagues at Weill Cornell Medical College and other centres. One of their most incredible discoveries was that, before cells detach from a primary tumour to begin their metastatic travels, proteins have already been released from the tumour. These are carried in circulating blood until they stick to a site, in effect tagging what will become landing points for wandering tumour cells. Imagine Paget's reaction to the evidence that secondary tumour sites are determined *before* any tumour cell sets foot outside the confines of the primary. The released proteins are chemical messengers – shades of Virchow's 'juice' – and they act like cellular parking lot attendants.

Tumours also release another signalling protein that acts in the bone marrow – the place where blood cells (red cells, white cells, etc.) are made from stem cells. This signal provokes the release of cells from the bone marrow, each carrying two protein markers on their surface: one sticks to the pre-marked landing site, the other hooks tumour cells from the circulation. It's a double-tagging process: the first messenger makes a sticky patch for bone marrow cells, released courtesy of another messenger, and these are the target for tumour cells. It's molecular Velcro: David Lyden calls it 'cellular bookmarking'.

The most recent instalments of this saga have revealed that the protein messengers that do the site-tagging are borne through the circulation by small sacs – mini cells called exosomes – that, amazingly, carry about 1,000 different types of protein. Specific protein tags home in on specific addresses found on different organs. You could think of the whole thing as a subway system with stations (lung, liver, etc.), each having its own code, recognized by exosome-borne proteins as they arrive on the circulatory train. The family of proteins that carry diverse addresses are called integrins and there are a couple of dozen varieties – more than enough to specify the major organs.

Controlling Metastatic Take-Off

Once tumour cells have 'colonized' a second home it is fortunate that, rather than taking off and expanding as a secondary tumour, they tend to do precisely the opposite, going into a state of harmless hibernation as the dormant growths we noted earlier. Lyden's group modelled this 'pre-metastatic niche' for human breast cancer cells in mice and identified two proteins released by nearby blood vessels that essentially work as a switch: one suppresses metastatic growth and it is not until the level of the other rises that the micro-metastases start to grow and become fully malignant. It is possible that these results in a mouse model system may not reflect what goes on in humans, but the 'switch' proteins are plausible candidates and they offer targets for investigation of their therapeutic potential.

Breaking the Barrier

Exosome messengers have one other very important trick in their armoury: they can cross the blood–brain barrier – the layer of (endothelial) cells that encloses the brain and controls the types of molecules that can move to and from circulating blood. This can be shown by pre-treating brain slices in dishes with exosomes from human breast cancer metastatic cells known to spread preferentially to different tissues (brain, lung or bone). Bathing the slices with brain-seeking exosomes caused a huge increase in the number of cancer cells attaching to the brain tissue. A specific protein has been identi-fied as a marker for these exosomes that is not present in exosomes that recruit lung or bone metastatic cells.

This remarkable work has revealed how exosomes help wandering tumour cells to storm the blood–brain barrier. Immediately this opens the possibility of isolating exosomes from small samples of blood and screening them for proteins – that is, using them as a 'biomarker' for metastatic cancer targets. This approach has already identified dozens of cancer-associated protein markers from tumour cells, from the tumour microenvironment and from immune system cells. The analysis of these biomarkers is reported to give a sensitivity of 95 per cent and a specificity of 90 per cent for cancer detection, opening the possibility that this screening method could one day be used in the clinic.

The importance of these advances is amplified by an unexpected feature of brain tumours. You might guess that they would start in the brain, but it turns out that most do nothing of the sort. The vast majority (about 90 per cent) are secondary cancers – that is, they arise when tumour cells spread from another part of the body, commonly breast or lung. In other words, most brain tumours are metastases – and they are mighty important. About 24,000 people in the USA discover they have these abnormal growths every year and they cause about 18,000 deaths. The rates are much the same in the UK, where deaths from brain and related tumours number just over 5,000.

Shooting the Messenger

The picture that has come into focus is that for disseminating tumours their exosomes are the scouts who do the foot-slogging: the protein signatures on the surface of these small, tumour-secreted packages home in on postcodes/zipcodes that define a desirable locale for metastatic spread. The obvious question is: 'If exosomes are critical in defining metastatic sites, can you block their action – and what happens when you do?' In preliminary experiments Ayuko Hoshino, David Lyden and colleagues have shown that either genetically knocking out specific integrins or blocking their capacity to stick to their targets (e.g., with a specific antibody) significantly reduces exosome adhesion, thereby blocking pre-metastatic niche formation and liver metastasis.

Metastatic behaviour is highly variable between different types of cancer. Some have usually spread by the time they are detected (lung, pancreatic), whereas breast and prostate tumours generally have not. Equally bemusing is the range of secondary targets: prostate cancer cells commonly home in on bone, whereas bone and muscle tumours often spread to the lungs. More promiscuous are triple-negative breast cancer, skin melanoma and tumours originating in the lung and kidney that can alight at multiple sites. About 10 per cent of diagnosed cancers are secondary growths with no detectable primary – they're 'cancers of unknown primary'.

Through these very recent discoveries our ignorance of how tumours spread is, at long last, beginning to be chipped away, and we'll return to this story in the next chapter when we look into the future of cancer detection and response to therapy.

Infection by Oncogenic Viruses

Although sometimes floated as a goal, the concept of a general cancer 'vaccine' is not realistic. Vaccines work against diseases caused by a single event – an infection of some sort – and hence are unsuitable for mutation-driven cancers. Nevertheless, effective prophylactic vaccines are available against some DNA tumour viruses that cause cancers subsequent to infection. These are designed to prevent the initiation of cancer and work in a similar manner to traditional vaccines by activating an immune response in the recipient, the antibodies thus produced acting to block infection. Prophylactic vaccines are available for the DNA human papillomaviruses (HPVs) and for hepatitis B virus (HBV). Chronic HBV infection can lead to hepatitis, liver cirrhosis and hepatocellular carcinoma, one of the five major cancers in the world. Recombivax HB and Engerix-B are FDA-approved vaccines for HBV. Approximately 70 per cent of cervical cancers and about 5 per cent of *all* cancers worldwide are caused by HPVs, and one of the great triumphs of cancer science has been the development of vaccines (Cervarix and Gardasil) that appear to give almost complete protection against infection by the tumour-promoting HPVs. The vaccines are artificially synthesized, non-infectious, virus-like particles that induce a strong immune response. The antibodies produced block binding of HPV to the cells lining the cervix (the epithelium) and hence prevent infection. These vaccines, however, have no effect as therapeutic vaccines, they can only prevent infection: once infection has occurred they are of no use.

Therapeutic Vaccines

A second type of vaccine is designed to boost the immune response by vaccination with fragments from proteins expressed specifically on the surface of tumour cells, a considerable number of which have now been identified. Progress in this field has been slow, but sipuleucel-T (Provenge) was approved by the FDA in 2010 as the first therapeutic vaccine. The immune response it provokes has prolonged the lives of patients with advanced-stage hormone-resistant prostate cancer. Two other vaccines have FDA approval: bacillus Calmettle-Guerin (TheraCys) for bladder cancer and T-VEC that we mentioned as the first oncolytic virus to be used clinically. In general, despite some positive results, therapeutic vaccines have been disappointing.

Tumour Agnostic Drugs

A new field of chemotherapy has seen the development of agents that target cancers independent of tissue or specific mutations. One such drug, pembrolizumab, is an antibody that blocks a checkpoint built into the immune response to prevent excessive activity – that is, a misdirected response that occurs when the system attacks cells of its own body. Larotrectinib is the most recent member of this class, being approved by the FDA in 2018 and for use in Europe in 2019. It is the first drug to be specifically approved to treat *any* cancer, hence the approval is 'tissue agnostic'. Larotrectinib is actually a tyrosine kinase inhibitor, some of which we've already met (e.g., gefitinib and erlotinib), but it blocks signalling from a very specific family of receptors – the three types of tropomyosin kinase receptors – that can drive a number of cell responses including avoidance of cell death. Although approved for use, clinical trials evaluating the drug are ongoing.

This approach to regulation should relieve a serious stumbling block for drug approval in that all non-targeted chemotherapeutic agents are 'tissue agnostic', but historically they have been approved for use in one cancer with extension for use in different types requiring new trials. A good example of this problem is gemcitabine, approved in the UK in 1995 and by the FDA in 1996 for pancreatic cancers. In 1998 the FDA extended its use to non-small-cell lung cancer and in 2004 it was approved for metastatic breast cancer in combination with paclitaxel.

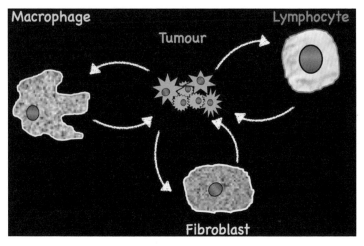

Figure 9.3   The tumour neighbourhood. Two-way communication between host cells and tumour cells. (A black and white version of this figure will appear in some formats. For the colour version, please refer to the plate section.)

The Tumour Microenvironment

Implicit in Paget's 'seed and soil' concept was the importance of the region surrounding a tumour mass, but only in the last two decades has it been possible to track the cellular comings and goings in the vicinity of tumours. Attracted initially by signals released from tumour cells, a variety of immune system cells and fibroblasts migrate to the neighbourhood, promoting angio-genesis and an immune response (Figure 9.3). This unseen world, now called the tumour microenvironment, is a cellular cloud composed of a variety of normal host cells that infiltrate and interact closely with the cancer cells. Over time a two-way communication develops, with messenger proteins shuttling between cells of the tumour and adjacent cells.

A Serious Case of Corruption

The cells of the immune system that are attracted to the site of a growing tumour are a very mixed bunch that includes white blood cells, including macrophages (that engulf foreign cells and material) and lymphocytes, of which the main types are B cells (that make antibodies) and an assortment of T cells including helper T cells (that direct the actions of other immune cells) and killer T cells (or cytotoxic T cells) that destroy cancer cells and virally infected cells. The immune system initially responds to a tumour by trying to eliminate the abnormal growth but, in an extraordinary transformation, when tumour cells manage to evade this defence the recruited cells change sides so to speak, switching to releasing signals that actively support tumour growth, angiogenesis and metastasis.

Attention has therefore focused on lifting the veil to reveal precisely who's doing what to whom in the tumour milieu. A particularly surprising finding has come from implanting human tumour cells into mice (the mice are 'immunocompromised' so they don't reject the human cells), which has revealed fibroblasts as a particularly potent driver of tumour growth and metastasis. These cells make the molecular scaffold that gives structure and shape to tissues and organs in animals, and we've already met them in the form of cancer-associated fibroblasts that release leptin to promote the growth and invasion of breast cancer cells. It's another example of nature using what is to hand – cells with a rather mundane function also playing an important role in cancer progression.

We don't know how effective the immune system is at eliminating cancers, but we do know that it is another example of a biological balancing act: it is activated to remove foreign organisms but over-activation results in auto-immunity when the system attacks normal tissues as if they were foreign organisms, examples being type 1 diabetes and rheumatoid arthritis.

The system is controlled by stimulatory and inhibitory checkpoints, much in the way that the cell cycle is regulated, and a dozen or so molecules have been identified as regulators of each type of checkpoint. The concept that it might be possible to interfere in the natural balance between stimulation and inhibition to boost immune activity in the tumour microenvironment, thereby using the host defence system to increase the efficiency of tumour elimination, is the basis of immunotherapy, and we will take up the advancing story of this exciting field in the next chapter.

10 The Road to Utopia?

In this final chapter we review briefly the current chemotherapy picture, highlighting the ingenious methods in development or in early trials that hold real promise. The cancer pathway has threaded its way through human history for four millennia and its course is marked by milestones of major advances that have saved many lives and offered much hope. It is also littered with the skeletons of failed experiments and dashed optimism. However, the present vista is breathtaking in that, as never before, the sciences of physics, chemistry and mathematics have converged on medicine to offer a bewildering cancer cocktail. Even with the warnings of history ringing in our ears, it seems reasonable to predict that the next 20 years will, at last, bring us to a point where we can regard most cancers as controllable. In consuming the elixir that follows we can be sure that some parts will fade to oblivion but others will surely survive and prosper to benefit all mankind.

Cancer Therapy: Immunotherapy

At the end of the previous chapter we sketched the tumour locale that has given rise to perhaps the most publicized area of current cancer research, namely immunotherapy – the manipulation of the immune system to treat disease. Activation immunotherapies elicit or amplify the response while suppression immunotherapies do the reverse. Four main strategies have been used: (1) immunomodulators – using a range of signalling proteins (cytokines or chemokines, either recombinant, synthetic or natural, e.g., interleukins) and immunomodulatory drugs (e.g., thalidomide and its derivatives); (2) vaccines and oncolytic viruses; (3) checkpoint inhibitors – antibodies that take the 'brakes' off the immune system, releasing it to attack cancer cells; and (4) gene therapy. We discussed the first two groups earlier and we now take up the story with checkpoint inhibitors and gene therapy.

Checkpoint Inhibitors

We left the assembled cast of the tumour microenvironment at the end of the last chapter with the notion that it might be possible to shift the balance of the immune system in favour of tumour cell elimination. The approach known as 'checkpoint blockade' uses agents that attenuate inhibitory pathways, releasing molecular brakes that normally prevent T cell hyperactivity and auto-immunity. It's a systemic method – drugs are administered that diffuse throughout the body to find their targets. There's nothing new about this general idea. Over 100 years ago the New York surgeon William Coley noticed that tumours sometimes vanished in patients with post-operative bacterial infections. He grew bugs in the laboratory and injected extracts into solid tumours, whereupon about 1 in 10 of them regressed, with some patients remaining well for many years thereafter.

Coley must have realized that further progress would depend on characterizing individual tumours and their supporting galaxy of cell types and messengers so that the optimal targets can be selected, as evidenced by his words of 105 years ago writ large on the lab notice board: 'That only a few instead of the majority showed such brilliant results did not cause me to abandon the method, but only stimulated me to more earnest search for further improvements in the method.'

It took a while before 'further improvements' were forthcoming – until, in fact, James Allison and Tasuku Honjo discovered that blocking negative immune regulation could be a form of cancer therapy. Allison showed that the T cell protein CTLA-4 acts as a brake on T cell activity and that an antibody to CTLA-4 could disengage the T cell brake and unleash the immune system to attack cancer cells. Honjo identified another T cell surface protein, PD-1, that also inhibits T cells. Under prolonged T cell stimulation the cells can up-regulate

PD-1 levels and tumour cells themselves produce a surface protein, PD-L1, that interacts with PD-1. Their interaction inhibits T cell activity (Figure 10.1).

Clinical trials showed striking effects for both types of checkpoint blockade, particularly in patients with advanced melanoma. Thus far, however, checkpoint therapy against PD-1 has proven more effective, with notable benefits for lung cancer, renal cancer, lymphoma and melanoma. Pembrolizumab, an anti-PD-1 antibody, was approved by the FDA in 2014. Combination therapy, targeting both CTLA-4 and PD-1, has shown increased efficacy for treating melanoma, as has combining anti-CTLA-4 monoclonal antibodies with vemurafenib. Immune checkpoint inhibitors are one of the most exciting developments in chemotherapy, not least because they are designed to re-invigorate

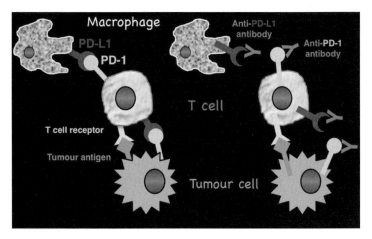

Figure 10.1   Checkpoint blockade of PD-1. Programmed cell death protein 1 (PD-1) is a cell surface receptor expressed on T cells and early-stage B cells. PD-1 binds to PD-L1 (and also to PD-L2). This contributes to the down-regulation of the immune system by preventing the activation of T cells, which in turn reduces autoimmunity and promotes self-tolerance. The inhibitory effect of PD-1 is accomplished through a dual mechanism of promoting apoptosis (programmed cell death) in antigen-specific T cells in lymph nodes while simultaneously reducing apoptosis in regulatory T cells (suppressor T cells). (A black and white version of this figure will appear in some formats. For the colour version, please refer to the plate section.)

the host's anti-tumour immune responses by disrupting inhibitory T cell signalling rather than acting directly on tumour cells – a concept reminiscent of anti-
angiogenic therapy. Nevertheless, successful outcomes thus far tend to occur
in only a small proportion of patients, around 10 per cent. This implies that, at
the moment, we have insufficient understanding of the interplay between
different cell populations in the tumour environment, a topic we'll return to
shortly, when we turn to MYC.

Gene Therapy

The notion that it might be possible to modify the genetic make-up of an
individual as a way of treating disease dates from the early 1970s. The idea is
to deliver nucleic acid (usually DNA) into cells to compensate for abnormal
genes using a carrier (vector), commonly a modified virus. It was first used in
humans in 1990 in two girls aged four and nine to correct an inherited disorder
called severe combined immunodeficiency that arises from a defective metabolic enzyme, adenosine deaminase. The field was shaken to its core when, in
1999, an 18-year-old American, Jesse Gelsinger, died four days after receiving an
injection of an adenovirus carrying a gene to compensate for another rare
metabolic disorder (ornithine transcarbamylase deficiency syndrome – OTCD).
The cause was a massive inflammatory response followed by major organ failure.
The enquiry revealed that the case had been handled in an extremely lax manner
and it's generally considered that Gelsinger's misfortune has benefited subsequent gene therapy patients through the introduction of rigorous controls.

A critical step in developing gene therapy for cancers tackled a weakness in
the immune system whereby it makes lymphocytes that kill tumour cells but
can't make enough – their protective effect is overwhelmed by the growing
cancer. Steven Rosenberg and colleagues at the National Cancer Institute,
Bethesda took small pieces of surgically removed tumours, grew them in the
lab, selected T cells with killing capacity, expanded them over a few weeks
and infused them back into the patient. This method, called adoptive cell
therapy, gives a hefty boost to the natural anti-tumour defence system and it
has been particularly effective for melanomas, including metastases.

The next refinement was to increase the efficiency of targeting tumour cells by
introducing new genes into the isolated T cells using a virus carrier, 'disabled'

so that it has none of its original disease-causing capacity but retains infectivity – it sticks to cells – with the required gene inserted. The method is an example of virotherapy. The gene used by Waseem Qasim and his colleagues at Great Ormond Street Hospital and University College London gave the T cells an artificial receptor that binds to CD19, present on the surface of acute lymphoblastic leukaemia cells. These receptors are known as chimeric T cell receptors or chimeric antigen receptors (CARs) – because they're made by fusing together several bits to make something that binds to the target 'antigen' – CD19. The first experiment treating B cell leukaemia was outstandingly successful and CAR-T cell therapy has been shown to be an effective novel therapeutic for leukaemias and lymphomas. In 2017 the FDA approved the first CAR-T cell therapy in the USA for acute lymphoblastic leukaemia and approved another gene therapy for an inherited form of vision loss. Over 2,600 gene therapy clinical trials have now been completed or are ongoing worldwide.

Figure 10.2 summarizes this gene-editing method. Also called 'genome editing' or 'genome editing with engineered nucleases', this form of genetic engineering removes or inserts sections of DNA, thereby modifying the genome. The 'cutting' is done by proteins, enzymes called nucleases – 'molecular scissors' – that snip both strands of DNA, creating double-strand breaks. This permits precise gene deletions or insertions and in-built systems then repair the DNA. The nucleases must target specific sites – identifiable by DNA sequence motifs. The leukaemia treatment used 'transcription activator-like effectors' but there are currently three other approaches to nuclease engineering, zinc finger nucleases, meganucleases and the CRISPR-Cas9 system. CRISPR-Cas9 has made the biggest headlines and it is now widely used for the insertion, removal or replacement of a specific gene(s) within a given genome and it has also been modified to make point mutations in DNA or RNA without making double-stranded DNA breaks – a technique called base editing.

What Is CRISPR-Cas9?

Viruses can attack bacteria just as they can human cells and, as part of their immune system, bugs have evolved a defence strategy by inserting short bits of viral DNA into their own genome so that they are primed to respond to

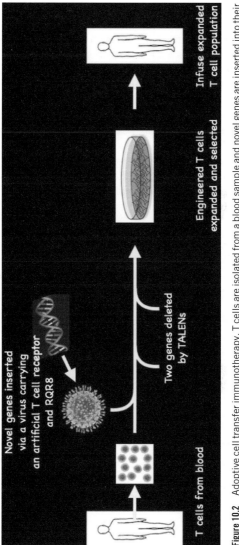

Figure 10.2 Adoptive cell transfer immunotherapy. T cells are isolated from a blood sample and novel genes are inserted into their DNA – an artificial receptor that binds to CD19 and RQR8 that encodes two proteins to help identification and selection of the modified cells. The method also uses gene editing by TALENs (transcription activator-like effectors) to delete two genes. The engineered T cells are expanded, selected and then infused into the patient. (A black and white version of this figure will appear in some formats. For the colour version, please refer to the plate section.)

Novel genes inserted via a virus carrying an artificial T cell receptor and RQR8

Two genes deleted by TALENS

T cells from blood

Engineered T cells expanded and selected

Infuse expanded T cell population

a subsequent infection by that virus. CRISPRs (standing for Clustered Regularly Interspaced Short Palindromic Repeats, pronounced *crispers*) refers to a family of short, repeated DNA sequences, each next to a 'spacer' sequence – one of the pieces of DNA taken from viruses that previously attacked the organism. Adjacent to CRISPR are genes encoding Cas9 enzymes that cut DNA, sometimes called 'nucleases' or 'nickases'. The CRISPR–spacer DNA is transcribed into RNA, to which Cas9 binds, and the complex roams the cell seeking a virus with genetic material that matches the CRISPR RNA. Once it sticks to the virus, Cas9 cleaves its DNA – end of virus. Thus, by binding to CRISPR RNA, Cas9 becomes an RNA-guided DNA cutter. The most recent advance in genome editing uses components from CRISPR systems together with other enzymes to insert point mutations into cellular DNA directly without making double-stranded DNA breaks.

The CRISPR system can be adapted to make the most versatile and precise method of genetic manipulation by designing guide RNA to bind to a target DNA sequence so that the molecular scissors (the Cas9 enzyme) cut the DNA at that sequence. The repair mechanisms of the cell either glue the DNA together or insert a novel stretch of DNA if that is delivered to the cell.

This permits editing of genes from any organism, and the development of CRISPR in the last decade or so has been one of the great advances in the life sciences, producing, for example, 'designer' immune cells with enhanced search-and-destroy capacities for tumours. Like all biological systems, however, CRISPR is not 100 per cent efficient, so there can be off-target effects and, as might be predicted, the act of cutting DNA activates TP53, which can be toxic to the cell.

The first steps have already been taken in direct administration of CRISPR-Cas9 gene therapy into the body. This was to treat a genetic condition that causes blindness (Leber's congenital amaurosis 10) and in the trial, named BRILLIANCE, the components of the gene-editing system – incorporated in the genome of a virus – are injected directly into the eye near photoreceptor cells. The aim of this approach is to remove a mutation in the gene (*CEP290*) causing this eye disorder. A further advance has used CRISPR to confront the problem of the failure of T cell therapy for solid tumours, notwithstanding its remarkable success in treating haematological malignancies. CAR-T cells were engineered to be resistant to PD-1 inhibition by disrupting three genes including PD-1 (*PDCD1*). The aim is to produce checkpoint blockade without using antibodies, and their direct injection to the brain has prolonged survival in mice with intracranial tumours.

CRISPR and related technologies are leading us into a new world in which Chinese scientists have already made the first CRISPR-edited human embryos and the first CRISPR-edited monkeys. The promise for cancer therapy is almost limitless.

Liquid Biopsy

We've noted that there are currently no serum biomarkers for cancer that approach an ideal specification and that screening using imaging is inadequate. Hundreds of potential biomarkers for the major cancers are being pursued and it may be that combinations of these will in time prove to be useful. The burgeoning field of metabolomics may also have a part to play, but at the moment the more promising approaches are liquid biopsies, breath analysis and developments in imaging.

Liquid biopsies use just a teaspoonful of blood from which circulating tumour cells, cell-free tumour DNA or exosomes can be recovered. Massively parallel sequencing is then used to detect tumour-specific alterations ranging from point mutations and small insertions and deletions to copy number variations, translocations and epigenetic changes. The mRNAs and proteins carried by exosomes can also be analysed. This non-invasive method permits tumour molecular profiling without the need for tumour tissue, and it offers the possibility of detecting cancers earlier than has hitherto been possible.

Liquid biopsy technology has been described as a transformative force in cancer care but it is, of course, limited by the amount of circulating tumour DNA released into the circulation. Not only does that vary between different cancers, but it will be at its lowest from early-stage cancers and from residual disease after surgery. The latter is important as the information could guide subsequent chemotherapy (as has been shown by mutant *BRAF* in melanoma) while early detection is one of the foremost challenges in cancer biology. Thus far the approach to mutation detection has been to sequence specific target sets (e.g., common cancer drivers) and to achieve accuracy by repeated sequencing. However, a recent, radical approach to this problem has supplanted depth of sequencing with breadth – that is, sequencing the entire genome rather than targeted regions to overcome the limitation of the low number of DNA fragments in plasma samples. This enhanced the sensitivity of circulating tumour DNA detection 100-fold, an advance that has exciting implications for liquid biopsy technology in early detection as an extension to its current use for monitoring response to treatment and the onset of resistance.

Breath Biopsy

At first pass it may sound fanciful to think of detecting cancer on the breath, but perhaps it shouldn't. After all, we're familiar with breathalysers that detect alcohol levels and, more generally, we all know that 'bad breath' isn't a good sign. For example, the smell of acetone on the breath can arise from type 1 diabetes, when the body increases its use of fat due to low insulin levels. The general point is that molecules released from cells can find their way into the lungs and emerge in the breath, and now they can be identified to find signatures indicative of disease.

Breath Biopsy® is an analytical platform using the most advanced chemistry (called field asymmetric ion mobility spectroscopy) that distinguishes molecules by how fast they move when driven through a gas by an electric field. Pioneered by the company Owlstone Medical, the idea is that exhaling into a mask permits breath-borne chemicals to be collected and analysed. More than 1,000 different compounds can be identified, including substances released by tumour cells and also those emanating from host cells in the tumour microenvironment. Breath Biopsy is undergoing trials in the hunt for biomarkers for the early detection of lung cancer.

Sponge on a String

In 1947 a consultant at St. Thomas' Hospital by the name of 'Pasty' Barrett successfully repaired a ruptured oesophagus – a surgical first for a hitherto fatal condition. Barrett noted that the cells lining the gullet sometimes change in appearance to resemble those found in the intestine. We now know that this change is caused by acid from the stomach being squeezed up into the oesophagus. Occasional regurgitation is called heartburn, but when it's persistent it becomes gastro-oesophageal reflux disease. In about 10 per cent of those cases sustained irritation by the stomach juices causes the change to what is now called Barrett's oesophagus. A few per cent of those with Barrett's oesophagus will get cancer of the oesophagus, the sixth most common cause of cancer-related death worldwide. Oesophageal cancer has become more common over the last 40 years, with men more prone to it than women, and it kills about 15,000 people in the USA each year and nearly 8,000 in the UK. Most cases aren't discovered until the disease has spread and it is then more or less untreatable. It's very bad news: the five-year survival figure is barely 15 per cent. Part of the problem is that the main sign is pain or difficulty in swallowing, often ignored until it is too late.

Until recently the only way of finding abnormal oesophageal tissue was by an endoscopy in which a tube with a camera is pushed down the throat, an unpleasant and expensive procedure. The desperate need for an easy, non-invasive test to screen for Barrett's oesophagus has recently been met by Cambridge oncologist Rebecca Fitzgerald, with a brilliantly simple development. The patient swallows a kind of 'teabag' on a string which is then pulled up from the stomach. The 'teabag' – called a Cytosponge – is a capsule about the size of a multivitamin pill containing a honeycomb sponge with a coating that dissolves in a few minutes when it reaches the stomach. When retrieved the sponge carries cells from the gullet lining (about half a million of them) that can then be analysed, and because cells are collected from the length of the gullet they give a complete picture rather than the local regions sampled in biopsies. The anticipation was that DNA sequencing of these cells would reveal the stages of oesophageal cancer development and hence whether a given case of Barrett's would or would not progress to cancer. Remarkably, however, whole-genome sequences from Barrett's and from oesophageal carcinoma showed that multiple mutations accumulate even in cells that are over-proliferating but look normal. As the condition progresses the range of mutations increases: in particular, regions of DNA are duplicated so that the genes therein are present in abnormal numbers. Typically there are 12,000 mutations per person with Barrett's oesophagus but without cancer and 18,000 mutations with cancer. Nevertheless, mutation patterns were identified that provide a highly accurate diagnosis of Barrett's oesophagus.

This impressive advance has given a clearer picture of the molecular basis of oesophageal cancer and, importantly, provided a screening method for early detection.

Epigenetics

The transcription of a gene from DNA into RNA – gene expression – is regulated through a variety of sequence elements that are distinct from the protein coding regions of genes. These include promoters, enhancers and super-enhancers. Genes thus become 'switched on' or 'switched off', as the jargon has it, by assemblies of proteins that bind to these elements and hence regulate the activity of enzymes that carry out transcription. In addition, gene expression is regulated though small chemical alterations to DNA that do not affect the sequence. This 'fine-tuning' is called epigenetic modulation – meaning any modification of DNA, *other* than in the sequence of bases, that affects how an organism develops or functions.

Perhaps the most familiar example of epigenetics in action comes in the shape of female tortoiseshell cats. Like humans, they have two X chromosomes, one of which is silenced in every cell. This can literally be seen in tortoiseshells because each X chromosome carries a fur-coloration gene (black and orange): inactivation of one chromosome produces the fur colour of the other. All this regulation of gene expression goes on in the two metres of DNA that is compressed into each nucleus of diameter six millionths of a metre (6 μm). To achieve this packaging, DNA strands are wrapped around scaffolding proteins to form a condensed structure called chromatin. Chromatin is further folded into higher orders of structure that give the characteristic shape of chromosomes. The chief proteins in chromatin are histones, which act as reels around which DNA wraps itself. Both histones and their entwined DNA are subject to epigenetic modification.

Two types of chemical alteration occur. The first tacks a small molecule (a methyl group, CH_3) either to A or C bases in DNA or to amino acids in histones and is carried out by methyltransferase enzymes. Histones can also be modified by addition of an acetyl group (CH_3CO) to lysine amino acids, the enzymes being histone acetyltransferases. These small chemical changes mediate epigenetic modulation of gene expression by altering the shape of chromatin and they can have profound effects on cell behaviour. Broadly speaking, methylation silences genes (switches them off). It's an essential part of normal development and many types of cancer have abnormal DNA methylation patterns, either over- or under-methylated, compared to normal tissue. One other feature of epigenetic modifications is that they are reversible, in contrast to mutations in DNA. Thus, de-methylases are enzymes that remove methyl groups; de-acetylases reverse histone acetylation.

The Dutch Famine

The importance of epigenetic changes in humans has been illustrated by the remarkable story of children conceived during the Dutch famine created by the German food blockade in the winter of 1944–1945 at the end of the Second World War. Some 20,000 died but, surprisingly given the dreadful conditions, records of healthcare were kept and these have been used to illuminate the subsequent lives of children who survived. A critical finding is that small babies resulted from malnourishment occurring only in the final months of pregnancy, whereas malnutrition during the first three months did not affect birth weight. An unexpected upshot, however, is that small babies remained small all their lives and had lower obesity rates, whereas those with normal birth weights who were malnourished only early in pregnancy were more prone to obesity. Recent DNA methylation analysis has shown that these effects could be passed on to the next generation. The gene involved is IGFII (insulin-like growth factor II) that is under-methylated relative to that in same-sex siblings who weren't prenatally exposed to famine. IGFII encodes a key protein in human growth and development, and its under-methylation suggests that its abnormal expression contributes to obesity.

Finding Cancer by Epigenetics

The finding that cancer cells often display abnormal DNA methylation patterns has stimulated epigenetic research into the 'methylscape' (methylation landscape) with a view to both diagnosis and therapeutic intervention. The development of computer programs that compare methylation profiles from solid tumours with those of healthy tissues has enabled the tumour origin to be identified from samples of circulating DNA. The program is called CancerLocator, and the initial study revealed different methylation patterns across the genomes of samples of lung, liver and breast cancer. This advance suggests a way round one of the problems with liquid biopsies, namely the inability to pinpoint the tissue of origin.

A further striking finding has been made by Matt Trau and his team at the University of Queensland, who discovered that changes in the methylation landscape affect the solubility of DNA and its adsorption by gold – how well it sticks. They exploited the differing stickiness of normal and cancer epigenomes using an indicator that changes colour depending on whether a sample extracted from blood contains DNA from tumour cells. This method can detect circulating free DNA from tumours within 10 minutes of taking a blood sample.

The impetus to develop these non-invasive tests is the requirement to detect cancers much earlier than can be achieved by current methods (mammography, etc.). The idea is that if we can detect cancers not weeks or months but perhaps years earlier, at that early stage they may be much more susceptible to drug treatments.

Liquid biopsies have already given useful information about patient response to treatment, but it will be a while before we can determine how far back any of these methods can push the detection frontier. In the meantime it would be interesting to apply these tests to age-grouped sets of normal individuals – because one would expect the frequency of cancer indication to be lower in younger people. From a scientific point of view it would be exciting if a significant proportion of 'positives' was detected in, say, 20–30-year-olds. Such a result would, however, raise some very tricky questions in terms of what, at the moment, should be done with those findings.

Epigenetic Drugs

As the importance of epigenetics in cancer has become recognized, much effort has gone into making epigenetic drugs (epidrugs), two classes of which have been approved by the FDA. An example is 5-azacitidine (AZA), which blocks the transfer of methyl groups to DNA. Because DNA methylation typically represses gene transcription, AZA can have the effect of reactivating aberrantly silenced tumour suppressor genes. It is used for some forms of leukaemia.

Another emerging class of epidrugs sticks to a family of proteins (BET proteins) that 'read' the epigenome by binding to acetylated lysines in histone proteins and turning on transcription. BET inhibitors have shown anti-cancer activity and are in late-stage clinical trials. One in particular, JQ1, blocks the expression of the *MYC* oncogene and we'll look at its potential shortly.

Nano-oncology

The use of gold nanoparticles to analyse DNA was our first encounter with the burgeoning field of cancer nanotechnology that aims to use nanoparticles to aid tumour imaging, biomarker detection and drug delivery. The term theranostics (therapeutics and diagnostics) is sometimes used to describe the approach that links diagnosis with treatment, increasingly mediated by nanotechnology. Materials in use include biodegradable polymeric nanoparticles, nanodiscs, highly branched molecules (called dendrimers), lipid vesicles (liposomes), metal nanoparticles, graphene and quantum dots (light-emitting crystals).

Nanoscale delivery systems have been developed for epidrugs – for example, linking AZA to polyethylene-vinyl acetate helps to protect the drug from degradation, enhances targeting specificity and minimizes adverse effects. A couple of other examples will give a flavour of the field.

Roboclot

These advances have given rise to the science of nanorobotics – making gadgets on a nanometre scale (10^{-9} metres or one-thousandth of a millionth of a metre). The gizmos themselves are nanorobots – nanobots to their friends – you can plonk a billion on the head of a pin. One example is a sort of molecular origami in which long strands of DNA (several thousand bases) are persuaded to fold into specific shapes by adding 'staples' – short bits of DNA that stick the strand to itself (Figure 10.3). When mixed the staples and DNA strands self-assemble in a single step. It's pretty amazing but it's driven by the simple concept of Watson–Crick base-pairing (A binds to T and G binds to C). Having made parcels of DNA the next step is to stick proteins to the DNA carrier that have a known target – for example, something on the

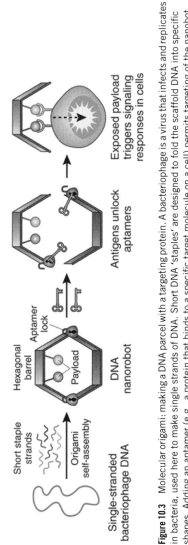

Figure 10.3 Molecular origami: making a DNA parcel with a targeting protein. A bacteriophage is a virus that infects and replicates in bacteria, used here to make single strands of DNA. Short DNA 'staples' are designed to fold the scaffold DNA into specific shapes. Adding an aptamer (e.g., a protein that binds to a specific target molecule on a cell) permits targeting of the nanobot. When it sticks to a cell the package opens and the molecular payload is released. (A black and white version of this figure will appear in some formats. For the colour version, please refer to the plate section.)

surface of a cell. The parcels can now be 'mailed' to that address in the body simply by injecting them into the bloodstream.

Hao Yan and colleagues from Arizona State University used a targeting protein that sticks to (endothelial) cells lining the walls of proliferating blood vessels (the target protein is called nucleolin). Generally these cells aren't proliferating so they don't make nucleolin – and the nanobots pass them by. But in growing tumours angiogenesis produces new vessels lined by endothelial cells to which Hao Yan's nanobots bind. The inclusion of a second protein that causes blood clotting (thrombin) yields nanobots that, in effect, cause thrombosis, inducing a blood clot to block the supply line to the tumour. It's also possible to tag these nanobots fluorescently to show that they do indeed target tumour blood vessels. Most critically the localized thrombosis caused by the released thrombin resulted in significant tumour cell death and marked increase in the survival of treated mice.

An extension of this nanobot approach has led to multifunctional nanoparticles for cancer immunotherapy. The principle is a kind of molecular Lego – making a series of separate bits and hooking them together. The trendy name is 'click chemistry', a term coined in 1998 by Barry Sharpless and colleagues at the Scripps Research Institute to describe reactions in which large, preformed molecules are linked to make even more complex multifunctional structures. One version of this approach uses lipid nanoparticles of polyethylene glycol with entrapped drugs. The nanoparticles are 'addressed' with an antibody that sticks to a receptor on tumour cells and blood vessels. Another polymer that causes the package to disintegrate inside cells is included, together with a chemical group that shows up brightly in MRI scans, the idea being to highlight where the nanoparticles go after injection. This type of construction has been used with sorafenib – commonly used to treat advanced liver cancers – as the entrapped drug. When injected into mice with liver tumours, these multifunctional nanoparticles do indeed home in on tumours and drastically reduce tumour growth.

This wonderfully clever chemistry will not cure cancers but it has already shown that hitting them in multiple ways can slow their growth. More potent drugs and further ingenuity will progressively extend this capacity.

3D Tumour Printing

Another stunning example of diverse scientific fields converging to produce novel anti-cancer strategies is the use of 3D printing to make a model system for the most common brain cancer, glioblastoma, for which the five-year survival rate is less than 5 per cent. The idea is to reproduce the tumour microenvironment by adding cancer cells to an extract of animal brain tissue together with endothelial cells (that line blood vessels) and gas-permeable silicone. This 'bio-ink' is printed onto a glass slide and grown for two weeks before testing combinations of candidate drugs to inform the treatment plan for the patient.

This tumour-on-a-chip method promises to be a significant advance in customizing treatments for glioblastoma. What's more, its use will not be limited to brain tumours. However, as always with scientific progress, it's not the final deal. For one thing, this system cannot reproduce an immune response that we know to be a critical modulator of tumour progression. Even so, it represents a quite astonishing marriage of scientific approaches to the problem of cancer treatment.

Targeted Alpha-Particle Therapy

The first stage of clinical trials has started for a new form of radiotherapy that has been nicknamed a 'magic bullet'. The 'bullet' is an alpha particle emitted by a radioactive isotope attached to a monoclonal antibody that binds to specific molecules on tumour cells. This form of radiation has a very short range but is lethal to nearby cells. The initial trial is for acute myelogenous leukaemia and, because these bullets reach their target within 10 minutes of administration, the strategy seems particularly suitable for blood-borne cancers.

Synthetic Lethality

When defects in a combination of two or more genes cause cell death but deficiency in only one allows the cell to survive, it's called synthetic lethality. Consider two separate signal pathways essential for the progression of a tumour, one resulting from a mutation that activates an oncoprotein (e.g.,

RAS), the other from the loss of a tumour suppressor. A drug that blocks the RAS pathway should block cancer development.

An example occurs in the two distinct pathways that repair damaged DNA. One normally repairs breaks in DNA using the PARP enzyme. When PARP is blocked by an inhibitor, breaks in DNA accumulate. Breaks can be repaired by a second pathway involving BRCA, whereupon the cell survives. However, in cancer cells with mutant BRCA this pathway cannot work. Hence, when a drug blocks PARP neither pathway can work. In effect, the inhibitor selectively kills cancer cells with BRCA mutations (Figure 10.4).

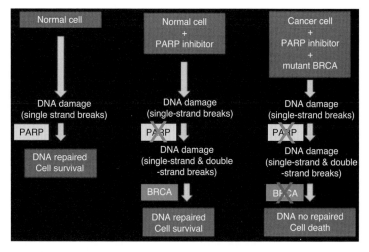

Figure 10.4 Synthetic lethality. If there are two distinct signalling pathways in a cell, each of which can be blocked without harming the cell but where simultaneous inhibition kills the cell, the effect is called synthetic lethality. The enzyme PARP (poly (ADP-ribose) polymerase 1) normally repairs single-strand DNA breaks. When this pathway is blocked by PARP inhibitors both single-strand and double-strand DNA breaks accumulate. If cells have normal BRCA it acts in a second pathway to repair DNA and the cell survives. However, in cancer cells with mutant BRCA this pathway is impaired. The use of PARP inhibitors means that neither pathway can work and the inhibitors, in effect, selectively target and kill cancer cells with BRCA mutations. (A black and white version of this figure will appear in some formats. For the colour version, please refer to the plate section.)

Clinical trials of several related PARP inhibitors, including olaparib, showed significant effects in ovarian cancer, and its use has been extended to prostate and pancreatic cancers with BRCA mutations. Enlarging the scope of a drug in this fashion is an example of the impact of cancer genomics on drug development, where finding an unexpected mutation (in BRCA in prostate and pancreatic cancers) offers the possibility of therapy with agents originally designed to treat breast cancer.

Personal versus Impersonal Medicine

The term 'personalized medicine' is very much in vogue but it's confusing because it can be taken to mean unique treatments designed for each individual. 'Precision medicine', often described as 'tailoring medical treatment to the individual characteristics of each patient' is not much better. To get things straight, let us say that the two expressions mean the same thing: genetic profiling of tumours to identify targetable abnormalities. Thus, vemurafenib blockade of mutant BRAF is an example. It's 'personal' in that it treats one specific mutation – but it's not 'personal' because anyone with that mutation is a potential candidate. Thus, the concept of personal medicine is the identification of features of a cancer that can be targeted by available treatments.

Before we leave 'personal' targets we should mention one further example that may be about to bear fruit: RAS proteins – molecular switches that form major nodes in the cellular network. *RAS* was the first human oncogene to be identified – in 1982 – and it's mutated in about 20 per cent of human tumours, permanently activating the MAPK pathway (see Chapter 6), an incentive for many ingenious efforts to 'drug' RAS. It's been a sobering story as RAS has emerged as one of the best examples of the paradox of cancer. On the one hand it seems startlingly simple but on the other it's been impenetrably complex, as illustrated by the huge amount of effort that has gone into blocking the action of mutant forms with no result. However, over the last 10 years that has begun to change and now at least five agents that modulate the KRAS member of the family are in clinical trials.

A promising example has come from Jude Canon and colleagues at Amgen Research, in the shape of a small molecule, AMG 510, that interacts with mutant KRAS to block GTP binding. The switch remains 'off' and the cancer-promoting activity of mutant KRAS is inhibited. Human pancreatic tumours in mice regressed completely in response to AMG 510 when combined with immunotherapy (an antibody against anti-PD1), and in a small trial of patients with non-small-cell lung cancer AMG 510 inhibited tumour growth. So maybe at long last the enigma of RAS is being prised open. The efficacy of AMG 510 against lung and pancreatic cancers is particularly heartening as there remains little in the way of therapeutic options for these tumours.

Notwithstanding recent advances and much publicity, personal medicine is yet to transform cancer therapy, prompting the thought that it might be productive to look instead at what cancers have in common, not least because resistance to drugs is a major problem with all forms of therapy, arising from the limitless flexibility of cellular control processes. The most obvious contender is MYC, a transcription factor and master regulator orchestrating the expression of a large number of genes (Chapter 6) that, collectively, initiate and maintain the growth of tumours as well as sustaining their microenvironment. We've noted that in most cancers control of MYC is lost and it's made in excessive amounts – up to 100 times the normal level. Because MYC appears to be a focus of many signalling pathways as they converge on the nucleus, it would appear to offer a 'pan-cancer' target, inhibition of which would block the growth of many cancers, regardless of tissue of origin and driver combinations (Figure 10.5). It is therefore an example of what has been called 'oncogene addiction', the dependence of continued cancer growth on one specific activity.

Genetic manipulation of MYC in mice has shown that reduction in MYC level causes rapid cessation of growth and regression of lung tumours driven by oncogenic RAS. Similar responses have been obtained in mice with small molecules that block MYC, but at the moment there are no clinically approved MYC inhibitors.

The Next Genomic Era

Since 2003 when the first complete human genome sequence was obtained, sequencing technology has transformed to the extent that genomes are now being sequenced on an industrial scale, with multiple coverage of all the

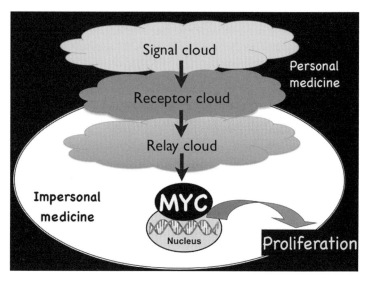

Figure 10.5 The central axis of cancer signalling. Many signalling pathways contribute to the tumour cell phenotype, but essentially they converge on a core of four tumour suppressors (ARF, p53, INK4s and RB1) and two oncogenes (*RAS* and *MYC*). These normally control cell division and ensure that damaged cells either become senescent or undergo apoptosis. Loss of function of any of the tumour suppressors or oncogenic activation of RAS or MYC is therefore a major step towards tumour initiation. (A black and white version of this figure will appear in some formats. For the colour version, please refer to the plate section.)

major cancers, as we noted in Chapter 4. Back then many would have supposed that, once we knew the most prominent cancer mutations, the power of protein chemistry would generate specific drugs and the era of personal medicine would have arrived. Well, it hasn't quite worked out like that, albeit that the revelations have been breathtaking and, as evidence of the power of modern sequencing, one could do no more than cite the Earth BioGenome Project that has been described as a moonshot for biology that aims to sequence, catalogue and characterize the genomes of *all* eukaryotes on Earth over a period of 10 years. Staggering though that is, as always in science great advances uncover new problems and questions. An obvious difficulty arising from the Pan-Cancer Analysis is that, of the 705 mutations that were repeatedly detected in cancer genomes, about 100 do not affect protein sequences and not only is it hard to pin down what they actually do, but restoring their normal activity would require a CRISPR-type engineering approach. A further problem is that the comprehensive genetic data are not matched to records of patient history, a limitation in tailoring therapy to individual patients. The UK's 100,000 Genomes Project is aiming to match DNA sequences with clinical information, and the next generation of cancer-genome sequencing projects will need to do likewise but this is not a trivial problem.

Even more of a challenge is the realization that cancer has turned out to be more complicated than anyone suspected. Thus, cancers that are clinically similar can contain very different sets of mutations – and not merely are no two cancers genetically the same but even adjacent cells within a tumour differ. Perhaps most sobering of all, Peter Campbell and colleagues at the Sanger Institute carried out a sequence analysis of samples of *normal* human skin taken from eyelids and found the number of mutations was similar to that in many cancers! Yet more amazing, the mutated genes included most of the key 'drivers' of a major type of skin cancer. The outer layers of normal skin were thus revealed as 'a battlefield of hundreds of competing mutant clones in every square centimetre of skin'. The obvious question, then, is 'Why isn't this tissue cancerous?', to which we have no answer. A plausible guess is that these cells either have a kind of master 'off switch' that suppresses potent driver combinations or they need a further 'on switch'. There's no evidence for either of these, nor is it clear whether other cell types can show this kind of restraint.

And there's another troubling point. Many cancer drugs are designed to target driver mutations and thus to kill the carrier cells. But, if these mutations can crop up in normal cells, any such 'cancer-specific' drugs might cause a good deal of what the military term 'collateral damage'.

Cancer Mosaics

A number of studies have thrown the might of DNA sequencing not merely at whole tumours but at small regions therein and at secondary growths, the idea being to chip out little bits from different regions of primaries and metastases. The picture that has emerged is of different regions having unique sequences – different 'DNA signatures'. In other words, although tumours start from a common ancestor, they diverge as they grow. So you can think of them like evolving species and draw family trees, as we did in Chapter 1, showing how long it was since you were a chimp or a ladybird.

It's even more complicated than that because even these sequences are averages – the predominant mutation pattern – because a small piece of tissue, say 1 mm in diameter, contains billions of cells, each with individual variations in its DNA sequence. The molecular complexity, even within one tumour, is utterly mind boggling, but there are two good reasons why we should make an effort to come to terms with it. The first is that, although cancers are an aberration, the cellular variety they embody is a wonder to behold and the adaptability – mutability if you like – that they reveal gives us a new vantage point from which to contemplate the breathtaking wonder of the natural world. The second is more prosaic: if we're going to reduce cancers to readily treatable conditions, we must understand fully the nature of the challenge – what makes a cancer cell tick. However stupefying the picture, that is what we are tackling and the more we know the more rational we can be in dealing with it.

Breast Cancer Mutational Signatures

The genetic dissection of breast cancer has extended this picture by looking at samples from multifocal disease ('foci' being small clumps of tumour cells). As expected, individual foci are related to their neighbours, but also have many 'private mutations' – ones found only in a single family, these being individual foci arising from a common ancestor. You could think of them as a cellular diaspora – a localized spreading or a kind of metastasis within the primary. A striking way of displaying these results is in 'coxcomb' plots – a variant of pie charts in which the colours and wedge sizes are different for every sample, representing – as they do – mutational diversity. It's a very graphic way of presenting information, devised by Florence Nightingale to show the causes of hospital deaths during the Crimean War.

Several mutational signatures for breast cancer have been teased from sequence data, notably the collection of 10 genomic driver-based sub-types from Carlos Caldas and colleagues at the Cambridge Breast Cancer Research

Unit. These advances are redefining breast cancer into sub-types classified by genetics rather than traditional pathology.

The Breast Tumour Microenvironment

We have noted the complexity of the tumour locale as many types of host cell accumulate and influence the tumour development. Startling evidence from Leeat Keren and colleagues at Stanford University has highlighted just how important this feature of solid tumours is going to be in the next phase of chemotherapy. They made a panel of antibodies that could pick out 36 different types of cell and tagged them for individual identification so that each cell type identified was given a false colour (pseudo colour) to show up the distribution. Adding these to slices of breast tumours yielded a set of mosaics so spectacularly beautiful it could pass as a group of miniatures by Georges Seurat. However, the cellular architecture revealed was seriously perturbing. In 40 sections of breast tumours, each from a different patient, no two sections appear at all similar. In other words, there were large differences in both the variety (type) and number of immune cells, notwithstanding the fact that the sections were all the same type of tumour – triple-negative breast cancer.

Nevertheless, even from this extraordinary variation it was possible to tease out some trends. Thus, for example, some tumours had high numbers of macrophages but low numbers of killer cells and intermediate numbers of other types of immune cells. Importantly, these patterns related to patient survival. Another notable finding was that proteins known to be important in controlling the immune response (e.g., PD-L1) appear in different cell types. For example, in tumour cells themselves in some patients but in immune cells in others.

These findings go some way to explaining why, thus far, immunotherapy has failed to produce beneficial effects in more than about 10 per cent of patients, the problem being that we just don't know enough about the tumour locale to determine the best approach. As we proceed, this type of cellular analysis will go hand in hand with genomics.

Contemplating the Portrait

It can be difficult when reviewing the efforts of mankind in the cancer field to suppress an image of a school report coming to mind – 'Tries hard but must do better.' With 10 million deaths a year and rising, even the most gung-ho boffin would be hard pushed to argue, although with a Gallic shrug they might observe that cancer research can feel like battling with a vast, Machiavellian organism devised by some malign, alien power in an unending struggle.

Understandable though that reaction may be, we have to resist. Many of us will have found ourselves muttering about our computers in human terms – they have minds of their own, they won't behave properly, etc. – even though it's you, the noodle, pressing the wrong button. There's a strong temptation to think of tumours in the same terms, but we have to keep reminding ourselves that cancer cells resemble computers in one way: they cannot think, for they are brainless. Like all cells, they depend on their insensate molecular machinery to propagate and migrate and when their hardware becomes corrupted, cellular behaviour follows suit. When we really feel up against it we should remind ourselves that the molecular gymnastics of Mother Nature have created the nine million or so different species on Earth and 100 times that number in the history of our planet – incomprehensible numbers representing quite staggering diversity. In taking on that force only a totally unfettered optimist would expect rapid progress.

In this chapter we have surveyed the major hurdles confronting cancer research and highlighted some of the astonishing ways in which every discipline of science is contributing to novel approaches for tumour detection, drug design and delivery and response monitoring. Large-scale DNA sequencing will continue to be at the forefront, given that the aim of precision medicine is to match patients to therapies on the basis of genomics. The recent huge advances in sequencing have submerged the cancer world in a tsunami of data, but rather than answering our prayers it has confronted us with yet more problems to surmount on the climb to achieving control of these diseases. The vast amounts of genetic data, the constantly shifting clonal populations of tumours and the corresponding complexity and fluidity of their environment, to say nothing of the relevant microbiome data, will all need to be taken into account. This integration will require sophisticated mathematical modelling, and we've noted the use of machine learning to identify mutation patterns common to different tumours. A subset of machine learning, deep learning, is being applied to scanning for patterns in breast cancer data to discern cellular changes that could inform drug discovery.

The vista opened by this amazing science encompasses the overall aim of categorizing patients to permit the rational design of therapies. However, we should record the words of caution from the Pan-Cancer Project:

> A major barrier to evidence-based implementation is the daunting heterogeneity of cancer chronicled in these papers, from tumour type to tumour type, from patient to patient, from clone to clone and from cell to cell. Building meaningful clinical predictors from genomic data can be achieved, but will require knowledge banks comprising tens of thousands of patients with comprehensive clinical characterization. As these sample sizes will be too large for any single funding agency, pharmaceutical company or health system, international collaboration and data sharing will be required.

It's now 60 years since the first drugs began to be used as a complement to surgery and radiotherapy for the treatment of cancers. The intervening period has seen dazzling triumphs with remission rates for some cancers that were formerly untreatable now at nearly 100 per cent and highly effective vaccines against some DNA tumour viruses. Innovative techniques are improving tumour detection and biomarkers isolated from blood can monitor response to therapy and are promising to reveal the earliest stages of cancer. Most dramatically of all, whole-genome sequencing has begun to change how we diagnose, classify and treat cancers and, through the detection of yet more mutations, is increasing the range of potential drug targets. However, there is a need to refine the selection of new drugs to reduce the number that are tested in clinical trials but never advance to FDA approval.

We're approaching the point where it will be possible to offer comprehensive screening to all new cancer patients, one benefit being that we'd stop giving people drugs that won't do any good because they won't hit the mutations carried by that individual. As we try to coerce cancers to our will we should rejoice in the tremendous successes that science has had in extending survival times for a substantial number of cancers such that many patients who decades ago would have received a death sentence now have to find other ways of dying. Nevertheless, the challenges remain daunting and there will be no 'quick fixes'. However, cancer biology will continue its steady advance to the benefit of mankind and all who follow the science and its application should be prepared to be amazed.

Concluding Remarks

The story of cancer began hundreds of millions of years ago and has continued in parallel with mammalian evolution. However, it only became a scourge in the twentieth century, mainly due to the doubling of the human lifespan. The era of molecular biology began in the second half of that century and this has unveiled much detail about the behaviour of cells and what goes wrong when they proliferate in an abnormal manner – the basic cause of cancers.

Astonishing technical advances led to the first sequence of human DNA in 2003. These have continued and vast databases now permit the identification of mutation signatures associated with most of the major types of cancer. This is changing the way cancers are classified and is facilitating the development of new drugs. While this has been viewed as ushering in the concept of personal medicine, it has also highlighted the problem of drug resistance – the extraordinary adaptability of cells in finding ways of neutralizing specific drugs, allowing them to continue to develop as tumours.

The tsunami of genomic data has also brought into focus many other areas of woeful ignorance. Thus, for example, the p53 tumour suppressor protein can switch on genes that make cells commit suicide – it's the 'guardian of the genome' – and its activity is lost in many cancers. However, despite over 50,000 papers being published on p53, we still don't really understand how the protein works.

One of the most recent advances is immunotherapy – the notion of giving the host immune system a boost to eliminate tumour cells. This is a promising

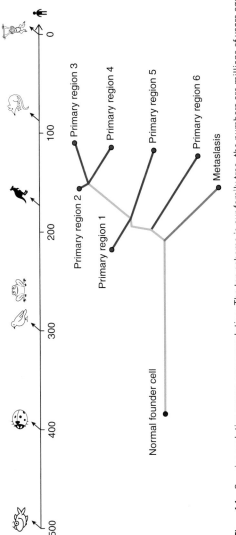

Figure 1.1 Species evolution versus cancer evolution. The top scheme is our family tree: the numbers are millions of years ago. The earth is about 4.5 billion years old, the first simple cells appeared around 3.8 billion years ago, and multicellular life started about 1,000 million years ago. The lower scheme represents the dissection of a tumour as an evolutionary tree. Six different regions of the same primary tumour and a secondary metastasis (M) are shown. The length of the lines between the branch points is proportional to the number of mutations picked up – thousands of them – on that stretch of the journey.

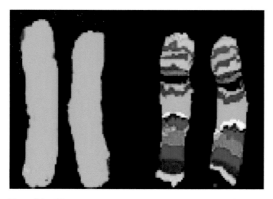

Figure 2.3 Chromosome rearrangements in cancer. Left: normal; right: tumour cell chromosomes.

(a)

Estimated number of new cases in 2018, worldwide, all cancers, both sexes, all ages

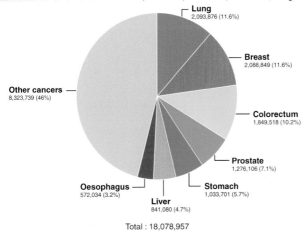

Lung
2,093,876 (11.6%)

Breast
2,088,849 (11.6%)

Colorectum
1,849,518 (10.2%)

Prostate
1,276,106 (7.1%)

Stomach
1,033,701 (5.7%)

Liver
841,080 (4.7%)

Oesophagus
572,034 (3.2%)

Other cancers
8,323,739 (46%)

Total : 18,078,957

(b)

Estimated number of deaths in 2018, worldwide, all cancers, both sexes, all ages

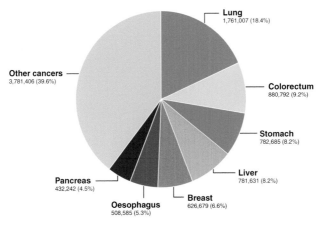

Lung
1,761,007 (18.4%)

Colorectum
880,792 (9.2%)

Stomach
782,685 (8.2%)

Liver
781,631 (8.2%)

Breast
626,679 (6.6%)

Oesophagus
508,585 (5.3%)

Pancreas
432,242 (4.5%)

Other cancers
3,781,406 (39.6%)

Total : 9,555,027

Figure 3.1 The 2018 global incidence (top) and mortality (bottom) for the top 10 major cancers (Globocan).

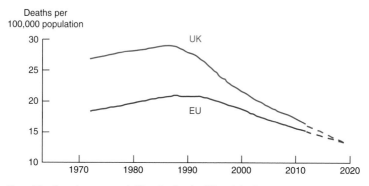

Figure 3.2 Breast cancer mortality rates for the UK and the European Union (age standardized). This shows the trend since 1970 and the predicted rates for 2019. The trends are similar for all cancer types taken together.

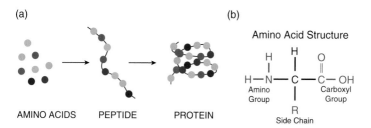

Figure 4.2 (a) Strings of amino acids make peptides and proteins. (b) Amino acids are molecules made up of an amine ($-NH_2$) and a carboxyl ($-COOH$) group attached to a carbon atom that is also linked to a side-chain (R group) specific to each of 20 different amino acids.

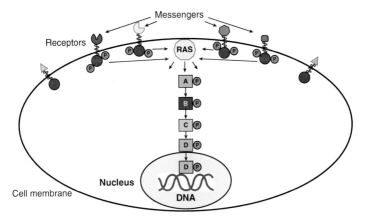

Figure 5.1 Cell signalling. Cells receive many signals from messengers that attach to receptor proteins spanning the outer membrane (RTKs). Activated receptors turn on relays of proteins. Receptors are shown as two blobs at either end of a wiggly line – the bit that crosses the membrane. RAS proteins are key nodes that transmit multiple signals. The shaded blocks represent a RAS-controlled pathway that 'talks' to the nucleus, switching on genes that drive proliferation. The arrows diverging from RAS indicate that it regulates many pathways controlling cell proliferation, cell differentiation, cell adhesion, cell death (apoptosis) and cell migration. Circled P = phosphate groups attached to the internal part of activated receptors; these act as launch pads for pathways that are activated by phosphorylation (e.g., A ⟹ B ⟹ C ⟹, etc.). The dramatic effects of phosphorylation on protein shapes has nothing to do with size – phosphate groups on proteins are like flies on elephants – it's the (negative) electrical charge they carry that distorts the shape and hence modulates the activity of a protein.

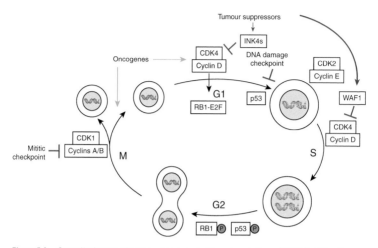

Figure 5.3 Overall picture of cell cycle control. Phosphorylation of RB1 in G1 releases transcriptional repression, enabling E2F proteins to activate key genes whose protein products drive cell cycle progression (*MYC*, *CDC2*, etc.). DNA damage activates p53: normal but not mutant p53 can then arrest the cell cycle and promote DNA repair after which p53 is switched off by phosphorylation. If DNA is not repaired p53 switches on genes that promote cell death. A further layer of control of CDKs and associated cyclins is by cyclin-dependent kinase inhibitors (CDIs) that include the *INK4* (inhibitor of CDK4) family and *WAF1*. RB1, p53 and the CDIs are known as tumour suppressors. Activating mutations in CDKs occur in some cancers, making them oncogenes.

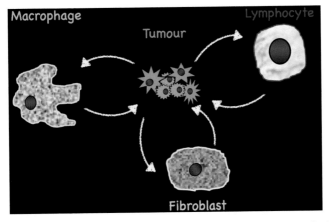

Figure 6.1 A mutational steeplechase leads to cancer. Of the tens of thousands of mutations that accumulate over time in a cancer cell, a handful of 'drivers' are critical (four are shown as danger [X] signs). Almost all mutations arise after birth (somatic), but about 1 in every 10 cancers start from a mutation present at birth (germline). Carriers are already one jump ahead and are much more likely to get cancer than those born with a normal set of genes. The rate at which mutations arise is increased by exposure to carcinogens (+) such as in tobacco smoke.

Figure 9.3 The tumour neighbourhood. Two-way communication between host cells and tumour cells.

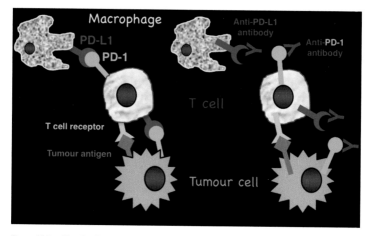

Figure 10.1 Checkpoint blockade of PD-1. Programmed cell death protein 1 (PD-1) is a cell surface receptor expressed on T cells and early-stage B cells. PD-1 binds to PD-L1 (and also to PD-L2). This contributes to the down-regulation of the immune system by preventing the activation of T cells, which in turn reduces autoimmunity and promotes self-tolerance. The inhibitory effect of PD-1 is accomplished through a dual mechanism of promoting apoptosis (programmed cell death) in antigen-specific T cells in lymph nodes while simultaneously reducing apoptosis in regulatory T cells (suppressor T cells).

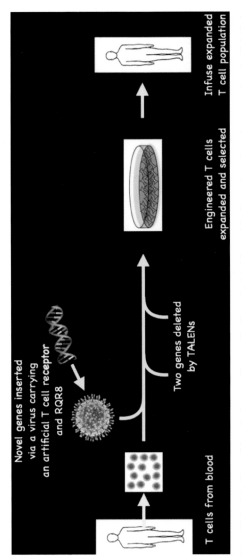

Figure 10.2 Adoptive cell transfer immunotherapy. T cells are isolated from a blood sample and novel genes are inserted into their DNA – an artificial receptor that binds to CD19 and RQR8 that encodes two proteins to help identification and selection of the modified cells. The method also uses gene editing by TALENs (transcription activator-like effectors) to delete two genes. The engineered T cells are expanded, selected and then infused into the patient.

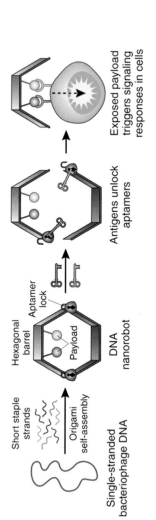

Figure 10.3 Molecular origami: making a DNA parcel with a targeting protein. A bacteriophage is a virus that infects and replicates in bacteria, used here to make single strands of DNA. Short DNA 'staples' are designed to fold the scaffold DNA into specific shapes. Adding an aptamer (e.g., a protein that binds to a specific target molecule on a cell) permits targeting of the nanobot. When it sticks to a cell the package opens and the molecular payload is released.

Short staple strands

Origami self-assembly

Single-stranded bacteriophage DNA

Hexagonal barrel

Aptamer lock

Payload

DNA nanorobot

Antigens unlock aptamers

Exposed payload triggers signaling responses in cells

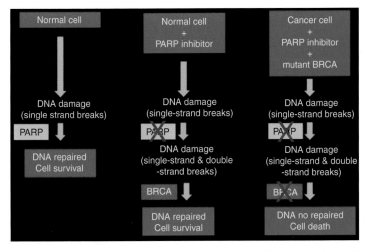

Figure 10.4 Synthetic lethality. If there are two distinct signalling pathways in a cell, each of which can be blocked without harming the cell but where simultaneous inhibition kills the cell, the effect is called synthetic lethality. The enzyme PARP (poly (ADP-ribose) polymerase 1) normally repairs single-strand DNA breaks. When this pathway is blocked by PARP inhibitors both single-strand and double-strand DNA breaks accumulate. If cells have normal BRCA it acts in a second pathway to repair DNA and the cell survives. However, in cancer cells with mutant BRCA this pathway is impaired. The use of PARP inhibitors means that neither pathway can work and the inhibitors, in effect, selectively target and kill cancer cells with BRCA mutations.

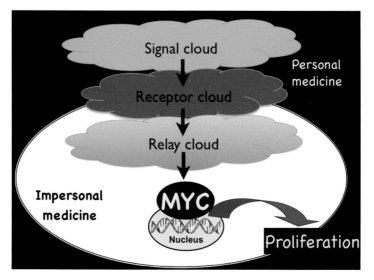

Figure 10.5 The central axis of cancer signalling. Many signalling pathways contribute to the tumour cell phenotype, but essentially they converge on a core of four tumour suppressors (ARF, p53, INK4s and RB1) and two oncogenes (*RAS* and *MYC*). These normally control cell division and ensure that damaged cells either become senescent or undergo apoptosis. Loss of function of any of the tumour suppressors or oncogenic activation of RAS or MYC is therefore a major step towards tumour initiation.

field, but it involves the constantly varying population of cells that migrates in and out of the tumour locale, and we know little of the cell types involved and the chemical signals that regulate their movements. Similarly, the processes that control the spread of cancer cells through the body – metastasis – are only just being revealed, and with them possible therapy targets, a critical area because secondary tumours cause over 90 per cent of cancer deaths.

Thus, our ignorance remains profound, but progress is being made on many fronts and it is reasonable to have a degree of cautious optimism for the future of mankind's struggle against these most complex and fascinating diseases.

Summary of Common Misunderstandings

Basic vocabulary. In science understanding the precise meaning of words is crucial and frequently encountered problems in cancer terminology are resolved in this book. An example is the word cancer itself: it means a growth, a mass (of cells) or a tumour – and these are used interchangeably. By 'cancer' a doctor means that a growth/mass/tumour has been specifically identified, either by histology (i.e., microscopic observation of the tissue) or by imaging (e.g., by MRI or PET).

Under-appreciation of risk factors. Despite many years of public information there remains a widespread inability to grasp the seriousness of, for example, smoking as a risk factor not only for many types of cancers but for a wide range of other diseases. A similar comment applies to poor eating habits.

Getting cancer will kill you sooner rather than later. Fifty years ago this was largely true but it is no longer. Survival rates for many cancers, including two of the most common, breast and prostate, have climbed to around 90 per cent and many who are diagnosed with cancer today will not die from it. There are, however, types of cancer that have not seen such improvements and the prognosis for them remains grave.

Cancer can be 'caught' either from others or by picking up some rogue gene from somewhere. No, it cannot, other than in the indirect sense that some cancers arise from viral or bacterial infections that can be transmitted, such as by sexual intercourse. No, it does not involve stray genes. It does involve changes in genes (mutations), but the genes involved are essential parts of our genetic make-up. They control the fundamental processes of cell growth – which is why abnormal versions can have such devastating effects.

Cancer can be cured. In a very restricted sense a number of cancers can be treated so that there is no evidence of their having been present and it is unlikely that they will recur. Unlikely, but not impossible. Thus, these cancers can be 'managed' – that is, kept under control. Other types of cancer remain essentially untreatable.

Cancer will eventually be eliminated. No, it won't, whatever triumphs medical science may achieve. Cancers are built into the workings of mammals. It's likely that almost all adult humans are carrying around in their tissues small, dormant tumours – most of which will not be a problem throughout a normal lifetime. However, estimates of mutation rates indicate that, if the lifespan of humans is extended to 140 years, all will have cancer.

Cancers are all the same. All cancers are different when their DNA sequences are analysed – that is, every cell differs from its neighbours. Many of these differences in DNA sequence don't matter, but some do and that's why there is no guarantee that any treatment will work nor that any remission will be permanent. The difference between types of cancer is important: high-grade sarcomas that have spread to distant parts of the body carry only a 15 per cent five-year survival rate, whereas basal cell carcinoma is usually a benign form of skin cancer that rarely spreads and can be readily treated.

Cancers are simply a mass of growing cells. It is only in the last few years that it has become clear that the tumour environment is in constant flux in terms of the movement of host cells that play roles in regulating tumour growth and dispersion. While this offers therapeutic options, it is not well understood and the concept is difficult to grasp – both for patients and for clinicians who may have to make life-altering treatment choices based on single biopsy samples that cannot convey temporal and special heterogeneity.

References

Chapter 1

Schwartz, R., and Schäffer, A.A. The evolution of tumour phylogenetics: principles and practice. *Nat Rev Genet*. 2017;18(4):213–229. doi: 10.1038/nrg.2016.170

Manchanda, R., Sun, L., Patel, S., et al. Economic evaluation of population-based *BRCA1/BRCA2* mutation testing across multiple countries and health systems. *Cancers*. 2020;12(7):1929.

National Cancer Institute. BRCA gene mutations: cancer risk and genetic testing. www.cancer.gov/about-cancer/causes-prevention/genetics/brca-fact-sheet#r2.

Ford, D., Easton, D.F., Stratton, M., et al. Genetic heterogeneity and penetrance analysis of the BRCA1 and BRCA2 genes in breast cancer families *Am J Hum Genet*. 1998;62(3):676–689. doi: 10.1086/301749

Ford, D., Easton, D.F., Bishop, D.T., Narod, S.A. and Goldgar, D.E. Risks of cancer in BRCA1-mutation carriers. *Lancet*. 1994;343(8899):692–5. doi: 10.1016/s0140-6736(94)91578-4

Kuchenbaecker, K.B., Hopper, J.L., Barnes, D.R., et al. Risks of breast, ovarian, and contralateral breast cancer for BRCA1 and BRCA2 mutation carriers. *JAMA*. 2017;317(23):2402–2416. doi: 10.1001/jama.2017.7112

Atchley, D.P., Albarracin, C.T., Lopez, A., et al. Clinical and pathologic characteristics of patients with BRCA-positive and BRCA-negative breast cancer. *J Clin Oncol*. 2008;26(26):4282–4288. doi: 10.1200/JCO.2008.16.6231

Chapter 2

Haridy, Y., Witzmann, F., Asbach, P., et al. Triassic cancer-osteosarcoma in a 240-million-year-old stem-turtle. *JAMA Oncol.* 2019;5(3):425–426. doi: 10.1001/jamaoncol.2018.6766

Chapter 3

Global Cancer Observatory. Home page. https://gco.iarc.fr/

NCRAS. Cancer e-Atlas. www.ncin.org.uk/cancer_information_tools/eatlas.

American Cancer Society. Cancer facts & figures 2020. www.cancer.org/research/cancer-facts-statistics/all-cancer-facts-figures/cancer-facts-figures-2020.html

American Cancer Society. Cancer facts & figures 2014. www.cancer.org/acs/groups/content/@research/documents/webcontent/acspc-042151.pdf.

US Census. Quick facts: United States. www.census.gov/quickfacts/fact/table/US/PST045218

Chapter 4

ICGC/TCGA Pan-Cancer Analysis of Whole Genomes Consortium, Campbell, P.J., Getz, G., et al. Pan-cancer analysis of whole genomes. *Nature.* 2020;578:82–93. doi: 10.1038/s41586-020-1969-6.

Meselson, M., and Stahl,F.W. The replication of DNA in *Escherichia coli. PNAS.* 1958;44(7):671–682. doi: 10.1073/pnas.44.7.671

Chapter 5

Nair, A., Chauhan, P., Saha, B. and Kubatzky, K.F. Conceptual evolution of cell signaling. *Int J Mol Sci.* 2019;20(13):3292. doi:10.3390/ijms20133292

Chapter 6

Caravagna, G., Giarratano, Y., Ramazzotti, D., et al. Detecting repeated cancer evolution from multi-region tumor sequencing data. *Nat Methods.* 2018;15(9):707–714. doi: 10.1038/s41592-018-0108-x

Chapter 7

zur Hausen, H. (2012). Red meat consumption and cancer: reasons to suspect involvement of bovine infectious factors in colorectal cancer. *Int J Cancer.* 130:2475–2483. doi: 10.1002/ijc.27413

Candelaria, P.V., Rampoldi, A., Harbuzariu, A. and Gonzalez-Perez, R.R. Leptin signaling and cancer chemoresistance: perspectives. *World J Clin Oncol.* 2017;8(2):106–119. doi: 10.5306/wjco.v8.i2.106

Alexandrov, L.B., Ju, Y.S., Haase, K., et al. Mutational signatures associated with tobacco smoking in human cancer. *Science.* 2016;354(6312):618–622. doi: 10.1126/science.aag0299.

Hijazi, K., Malyszko, B., Steiling, K. et al. Tobacco-related alterations in airway gene expression are rapidly reversed within weeks following smoking-cessation. *Sci. Rep.* 2019;9(6978). doi: 10.1038/s41598-019-43295-3

Yoshida, K., Gowers, K.H.C., Lee-Six, H., et al. Tobacco smoking and somatic mutations in human bronchial epithelium. *Nature.* 2020;578(7794):266–272. doi: 10.1038/s41586-020-1961-1.

Pleguezuelos-Manzano, C., Puschhof, J., Rosendahl Huber, A., et al. Mutational signature in colorectal cancer caused by genotoxic pks+ E. coli. *Nature.* 2020;580(7802):269–273. doi: 10.1038/s41586-020-2080-8.

Mortazavi, S.M.J., Ghiassi-Nejad, M. and Rezaiean, M. Cancer risk due to exposure to high levels of natural radon in the inhabitants of Ramsar, Iran. In *International Congress Series*, volume 1276, 2005, pp. 436–437, doi: 10.1016/j.ics.2004.12.012

Bryn Austin, S., Yu, K., Liu, S.H., Dong, F. and Tefft, N. Household expenditures on dietary supplements sold for weight loss, muscle building, and sexual function: Disproportionate burden by gender and income. *Prevent. Med. Rep.* 2017;6:236–241. doi: 10.1016/j.pmedr.2017.03.016

Lustig, R., Schmidt, L. and Brindis, C. The toxic truth about sugar. *Nature.* 2012;482:27–29. doi: 10.1038/482027a

Chapter 8

Roy Kishony's Laboratory at Harvard Medical School. www.youtube.com/watch?
v=plVk4NVIUh8

Hanahan, D., and Weinberg, R.A. Hallmarks of cancer: the next generation.
Cell.2011;144:646–674. doi: 10.1016/j.cell.2011.02.013

Wikipedia. Protein kinase inhibitor. https://en.wikipedia.org/wiki/
Protein_kinase_inhibitor.

Cancer Research UK. Breast screening. www.cancerresearchuk.org/about-cancer
/breast-cancer/getting-diagnosed/screening/breast-screening.

Luengo, A., Gui, D.Y. and Vander Heiden, M.G. Targeting metabolism for cancer
therapy. *Cell Chem Biol*. 2017;24(9):1161–1180. doi: 10.1016/j.
chembiol.2017.08.028

Mullard, A. FDA approves first-in-class cancer metabolism drug. *Nat Rev Drug
Discov*. 2017;16(593). doi: 10.1038/nrd.2017.174

Welti, J., Loges, S., Dimmeler, S. and Carmeliet, P. Recent molecular discoveries
in angiogenesis and antiangiogenic therapies in cancer. *J Clin Invest*. 2013;123
(8):3190–3200. doi: 10.1172/JCI70212

Tebbutt, N., Pedersen, M.W. and Johns, T.G. Targeting the ERBB family in
cancer: couples therapy. *Nat Rev Cancer*. 2013;13(9):663–673. doi:
10.1038/nrc3559

Kaplan, R.N., Riba, R.D., Zacharoulis, S., et al. VEGFR1-positive haematopoietic
bone marrow progenitors initiate the pre-metastatic niche. *Nature*. 2005;438
(7069):820–827. doi: 10.1038/nature04186

Ghajar, C.M., Peinado, H., Mori, H., et al. The perivascular niche regulates
breast tumour dormancy. *Nat Cell Biol*. 2013;15(7):807–817. doi: 10.1038/
ncb2767

Brabletz, T., Lyden, D., Steeg, P. et al. Roadblocks to translational advances
on metastasis research. *Nat Med*. 2013;19:1104–1109. doi: 10.1038/
nm.3327

Rodrigues, G., Hoshino, A., Kenific, C.M., et al. Tumour exosomal CEMIP protein promotes cancer cell colonization in brain metastasis. *Nat Cell Biol.* 2019;21 (11):1403–1412. doi: 10.1038/s41556-019-0404-4

Hoshino, A., Costa-Silva, B., Shen, T.L., et al. Tumour exosome integrins determine organotropic metastasis. *Nature.* 2015;527:329–335. doi: 10.1038/nature15756

Hoshino, A., Kim, H.S., Bojmar, L., et al. Extracellular vesicle and particle biomarkers define multiple human cancers. *Cell.* 2020;182:1–18. doi: 10.1016/j.cell.2020.07.009

Ruoslahti, E. RGD and other recognition sequences for integrins. *Annu Rev Cell Dev Biol.* 1996;12:697–715. doi: 10.1146/annurev.cellbio.12.1.697

Coley, W.B. The treatment of inoperable sarcoma by bacterial toxins (the mixed toxins of the *Streptococcus erysipelas* and the *Bacillus prodigiosus*). *Proc R Soc Med.* 1910;3:1–48.

Twyman-Saint Victor, C., Rech, A.J., Maity, A., et al. Radiation and dual checkpoint blockade activate non-redundant immune mechanisms in cancer. *Nature.* 2015;520(7547):373–377. doi: 10.1038/nature14292.

Wolchok, J.D., Kluger, H., Callahan, M.K., et al. Nivolumab plus ipilimumab in advanced melanoma. *N Eng J Med.* 2013;369:122–133. doi: 10.1056/NEJMoa1302369

Chapter 9

Biller-Andorno, N., and Jüni, P. Abolishing mammography screening programs? A view from the Swiss Medical Board. *New Eng J Med.* 2014;370:1965–1967.

Independent UK Panel on Breast Cancer Screening. The benefits and harms of breast cancer screening: an independent review. *Lancet.* 2012;380:1778–1786.

Gøtzsche, P.C., and Jørgensen, K.J. Screening for breast cancer with mammography. *Cochrane Database Syst Rev.* 2013;6: CD001877. doi: 10.1002/14651858.CD001877

Chapter 10

Explore Gene Therapy. History of gene therapy. www.exploregenetherapy.com/history-of-gene-replacement-therapy

Wikipedia. Gene therapy. https://en.wikipedia.org/wiki/Gene_therapy

Rosenberg, S.A., and Restifo, N.P. Adoptive cell transfer as personalized immuno-therapy for human cancer. *Science*. 2015;348(6230):62–68. doi: 10.1126/science.aaa4967

Choi, B.D., Yu, X., Castano, A.P. et al. CRISPR-Cas9 disruption of PD-1 enhances activity of universal EGFRvIII CAR T cells in a preclinical model of human glioblastoma. *J Immunother Cancer*. 2019;7(304). doi: 10.1186/s40425-019-0806-7

Zviran, A., Schulman, R.C., Shah, M., et al. Genome-wide cell-free DNA muta-tional integration enables ultra-sensitive cancer monitoring. *Nat Med*. 2020;26 (7):1114–1124. doi: 10.1038/s41591-020-0915-3

Aktas, B., Muller, V., Tewes, M., et al. Comparison of estrogen and progesterone receptor status of circulating tumor cells and the primary tumor in metastatic breast cancer patients. *Gynecol Oncol*. 2011;122(2):356–360. doi: 10.1016/j.ygyno.2011.04.039

Owlstone Medical. *Breath Biopsy: The Complete Guide*. Cambridge: Owlstone Medical.

Cambridge University. 'Pill on a string' could help spot early signs of cancer of the gullet. www.cam.ac.uk/research/news/pill-on-a-string-could-help-spot-early-signs-of-cancer-of-the-gullet#sthash.Ue0mqOdP.dpuf

Cambridge University. Cytosponge: Early detection for oesophageal cancer [video]. www.youtube.com/watch?v=iGqBu4C2ASg&feature=youtu.be.

Ross-Innes, C.S., Becq, J., Warren, A., et al. Whole-genome sequencing provides new insights into the clonal architecture of Barrett's esophagus and esophageal adenocarcinoma. *Nat Genet*. 2015;47:1038–1046. doi: 10.1038/ng.3357

Kang, S., Li, Q., Chen, Q., et al. CancerLocator: non-invasive cancer diagnosis and tissue-of-origin prediction using methylation profiles of cell-free DNA. *Genome Biol*. 2017;18:53. doi: 10.1186/s13059-017-1191-5.

Sina, A.A., Carrascosa, L.G., Liang, Z., et al. Epigenetically reprogrammed methy-lation landscape drives the DNA self-assembly and serves as a universal cancer biomarker. *Nat Commun*. 2018;9(1):4915. doi: 10.1038/s41467-018-07214-w

Fu, J., and Yan, H. Controlled drug release by a nanorobot. *Nat Biotechnol* 2012;30:407–408. doi: 10.1038/nbt.2206.

Li, S., Jiang, Q., Liu, S., et al. A DNA nanorobot functions as a cancer therapeutic in response to a molecular trigger in vivo. *Nat Biotechnol* 2018;36:258–264. doi: 10.1038/nbt.4071

Liu, Y., Feng, L., Liu, T., et al. Multifunctional pH-sensitive polymeric nanoparticles for theranostics evaluated experimentally in cancer. *Nanoscale*. 2014;6 (6):3231–3242. doi: 10.1039/c3nr05647c

Yi, H.G., Jeong, Y.H., Kim, Y., et al. A bioprinted human-glioblastoma-on-a-chip for the identification of patient-specific responses to chemoradiotherapy. *Nat Biomed Eng*. 2019;3:509–519. doi: 10.1038/s41551-019-0363-x

Gomez-Roman, N., and Chalmers, A.J. Patient-specific 3D-printed glioblastomas. *Nat Biomed Eng*. 2019;3:498–499. doi: 10.1038/s41551-019-0379-2

Canon, J., Rex, K., Saiki, A.Y., et al. The clinical KRAS(G12C) inhibitor AMG 510 drives anti-tumour immunity. *Nature*. 2019;575:217–223. doi: 10.1038/s41586-019-1694-1

Lou, K., Steri, V., Ge, A.Y., et al. KRASG12C inhibition produces a driver-limited state revealing collateral dependencies. *Sci Signal*. 2019;12(583):eaaw9450. doi: 10.1126/scisignal.aaw9450

Soucek, L., Whitfield, J., Martins, C.P., et al. Modelling Myc inhibition as a cancer therapy. *Nature*. 2008;455(7213):679–683. doi: 10.1038/nature07260

Martincorena, I., Roshan, A., Gerstung, M., et al. High burden and pervasive positive selection of somatic mutations in normal human skin. *Science*. 2015;348:880–886. doi: 10.1126/science.aaa6806

Ali, H.R., Rueda, O.M., Chin, S.F. et al. Genome-driven integrated classification of breast cancer validated in over 7,500 samples. *Genome Biol*. 2014;15(431). doi: 10.1186/s13059-014-0431-1

De Mattos-Arruda, L., Sammut, S.J., Ross, E.M., et al. The genomic and immune landscapes of lethal metastatic breast cancer. *Cell Rep*. 2019;27:2690–2708. doi: 10.1016/j.celrep.2019.04.098

Yates, L.R., Gerstung, M., Knappskog, S., et al. Subclonal diversification of primary breast cancer revealed by multiregion sequencing. *Nat Med.* 2015;21 (7):751–759. doi: 10.1038/nm.3886

Robinson, D., Van Allen, E.M., Wu, Y.M., et al. Integrative clinical genomics of advanced prostate cancer. *Cell.* 2015;161(5):1215–1228. doi: 10.1016/j. cell.2015.05.001

Keren, L., Bosse, M., Marquez, D., et al. Structured tumor-immune, microenvironment in triple negative breast cancer revealed by multiplexed ion beam imaging. *Cell.* 2018;174(6):1373–1387.e19. doi: 10.1016/j.cell.2018.08.039

Gerlinger, M., Rowan, A., Horswell, S., *et al.* Intratumor heterogeneity and branched evolution revealed by multiregion sequencing. *New Eng J Med* 2012;366:883–892. doi: 10.1056/NEJMoa1113205

Adams, S. DNA map offers hope on cancer treatments. *Telegraph*, 28 January 2013. www.telegraph.co.uk/health/healthnews/9832535/DNA-map-offers-hope-on-cancer-treatments.html

Crowther, M.D., Dolton, G., Legut, M., et al. Genome-wide CRISPR-Cas9 screening reveals ubiquitous T cell cancer targeting via the monomorphic MHC class I-related protein MR1. *Nat Immunol.* 2020;21(2):178–185. doi: 10.1038/s41590-019-0578-8

Nobel Prize Winners Whose Work Contributed to the Story of Cancer

Wilhelm Conrad Röntgen. Generation and detection of X-rays or Röntgen rays. 1901 Nobel Prize in Physics.

Marie Curie, Pierre Curie and Henri Becquerel. Radioactivity. 1903 Nobel Prize in Physics.

Fritz Haber. Ammonia-based fertilizer. 1918 Nobel Prize in Chemistry.

Frederick Gowland Hopkins and Christiaan Eijkman. Discovery of vitamins. 1929 Nobel Prize in Physiology or Medicine.

Otto Warburg. Cellular respiration. 1931 Nobel Prize in Physiology or Medicine.

Frederick Joliot and Irene Joliot-Curie. Artificial creation of new radioactive elements. 1935 Nobel Prize in Chemistry.

Ernest Orlando Lawrence. Invention of the cyclotron. 1939 Nobel Prize in Physics.

George De Hevesy. Development of radioactive isotopes as tracers in animals. 1943 Nobel Prize in Chemistry.

Otto Stern. The development of the molecular ray method and discovery of the magnetic moment of the proton. 1943 Nobel Prize in Physics.

Isidor Isaac Rabi. Discovery of nuclear magnetic resonance. 1944 Nobel Prize in Physics.

Hermann Joseph Muller. Discovery that mutations can be induced by X-rays. 1946 Nobel Prize in Physics.

Edward Mills Purcell and Felix Bloch. Development of new methods for nuclear magnetic precision measurements. 1952 Nobel Prize in Physics.

John Franklin Enders, Thomas Huckle Weller and Frederick Chapman Robbins. Discovery of the ability of poliomyelitis viruses to grow in cultures of various types of tissue. 1954 Nobel Prize in Physiology or Medicine.

André Frédéric Cournand, Werner Forssmann, Dickinson W. Richards. Discoveries concerning heart catheterization and pathological changes in the circulatory system. 1956 Nobel Prize in Physiology or Medicine.

Fred Sanger. First protein sequence. 1958 Nobel Prize in Chemistry.

Arthur Kornberg and Severo Ochoa. DNA polymerases. 1959 Nobel Prize in Physiology or Medicine.

James Watson, Francis Crick and Maurice Wilkins. Structure of DNA. 1962 Nobel Prize in Physiology or Medicine.

Max Perutz and John Kendrew. Structures of haemoglobin and myoglobin. 1962 Nobel Prize in Chemistry.

Dorothy Hodgkin. Structure of vitamin B12. 1964 Nobel Prize in Chemistry.

François Jacob, André Lwoff and Jacques Monod. Genetic control of enzyme and virus synthesis. 1965 Nobel Prize in Physiology or Medicine.

Charles Huggins and Peyton Rous. Chemotherapy. 1966 Nobel Prize in Physiology or Medicine.

Marshall Nirenberg, Har Gobind Khorana and Robert Holley. Breaking the genetic code. 1968 Nobel Prize in Physiology or Medicine.

Salvador Luria, Max Delbrück and Alfred Hershey. Discovery of the replication mechanism and the genetic structure of viruses. 1969 Nobel Prize in Physiology or Medicine.

Howard Temin, Renato Dulbecco and David Baltimore. Discovery of reverse transcriptase. 1975 Nobel Prize in Physiology or Medicine.

Werner Arber, Hamilton Smith and Daniel Nathans. Discovery of restriction endonucleases. 1978 Nobel Prize in Physiology or Medicine.

Godfrey Hounsfield and Allan McLeod Cormack. Computed tomography. 1979 Nobel Prize in Physiology or Medicine.

Paul Berg, Walter Gilbert and Fred Sanger. DNA sequencing. 1980 Nobel Prize in Chemistry.

Nicolaas Bloembergen, Arthur Leonard Schawlow and Kai Manne Börje Siegbahn. Development of laser spectroscopy. 1981 Nobel Prize in Physics.

Norman Foster Ramsey. Invention of the separated oscillatory field method. 1981 Nobel Prize in Physics.

Michael Bishop and Harold Varmus. Cellular origin of retroviral oncogenes. 1989 Nobel Prize in Physiology or Medicine.

Richard Robert Ernst. Development of Fourier transform nuclear magnetic resonance spectroscopy. 1991 Nobel Prize in Chemistry.

Kary Mullis and Michael Smith. Inventing the polymerase chain reaction. 1993 Nobel Prize in Chemistry.

Tim Hunt, Lee Hartwell and Paul Nurse. Cell cycle. 2001 Nobel Prize in Physiology or Medicine.

John Sulston, Bob Horvitz and Sydney Brenner. Cell-lineage tree. 2002 Nobel Prize in Physiology or Medicine.

Paul Christian Lauterbur and Peter Mansfield. Development of magnetic resonance imaging (MRI). 2003 Nobel Prize in Physiology or Medicine.

Venkatraman Ramakrishnan, Thomas Steitz and Ada Yonath. Detailed structure and mechanism of the ribosome. 2009 Nobel Prize in Chemistry.

James Allison and Tasuku Honjo. Cancer therapy by immune regulation. 2018 Nobel Prize in Physiology or Medicine.

Figure Credits

Figure 1.1   Adapted from Gerlinger, M. et al., Intratumor heterogeneity and branched evolution revealed by multiregion sequencing. New Eng J Med (2012); 366:883–892. doi:10.1056/NEJMoa1113205

Figure 2.1   (a) Image by Rainer Schoch, taken from https://commons.wiki media.org/wiki/File:Bild2_Ur-Schildkr%C3%B6te_Zeichnung .jpg (made available under a CC BY-SA 4.0 licence). (b) Image kindly contributed by Dr Richard Hesketh, University College Hospital.

Figure 2.2   Taken from https://en.wikipedia.org/wiki/Robert_Hooke#/medi a/File:RobertHookeMicrographia1665.jpg (image in the public domain).

Figure 2.3   Reproduced from Kuechler et al., Precise breakpoint characterization of the colon adenocarcinoma cell line HT-29 clone 19A by means of 24-color fluorescence in situ hybridization and multicolor banding, *Genes, Chromosomes & Cancer* 36, 207 (2003), DOI: 10.1002/gcc.10163. Reprinted with permission from John Wiley and Sons.

Figure 3.1   Based on data from the WHO Globocan website.

Figure 3.2   Adapted from Malvezzi, M. et al., (2019). European cancer mortality predictions for the year 2019 with focus on breast cancer. Annals of Oncology 30, 781–787. https://doi.org/10

.1093/annonc/mdz051 Reprinted with permission from Elsevier.

Figure 4.1   Reproduced from Watson, J., Crick, F. Molecular Structure of Nucleic Acids: A Structure for Deoxyribose Nucleic Acid. *Nature* 171, 737–738 (1953). https://doi.org/10.1038/17173 7a0. Reprinted with permission from Springer Nature.

Figure 4.3   Originally drawn by Thomas Shafee.

Figure 5.2   Originally drawn by Thomas Shafee.

Figure 5.3   Originally drawn by Thomas Shafee.

Figure 6.2   Originally drawn by Thomas Shafee.

Figure 6.3   Originally drawn by Thomas Shafee.

Figure 6.5   Originally drawn by Thomas Shafee.

Figure 7.1   Drawn by the author based on data from cruk.org.

Figure 9.1   Left-hand image taken from https://en.wikipedia.org/wiki/Wilh elm_R%C3%B6ntgen#/media/File:First_medical_X-ray_by_W ilhelm_R%C3%B6ntgen_of_his_wife_Anna_Bertha_Ludwig's_ hand_-_18951222.gif (image in the public domain). Right-hand image kindly contributed by Dr Richard Hesketh, University College Hospital.

Figure 9.2   Originally drawn by Thomas Shafee.

Figure 10.3   Reproduced from Fu, Y., Chao, J., Liu, H. et al. Programmed self-assembly of DNA origami nanoblocks into anisotropic higher-order nanopatterns. Chin. Sci. Bull. 58, 2646–2650 (2013). https://doi.org/10.1007/s11434-012-5530-3 Reprinted from *Nature Biotechnology* with permission of Springer Nature.

Index

Page numbers in *italic* refer to figures; those in **bold** refer to tables
Numbers are filed as spelled out